LANGWITH JUNCTION : THE LIFE AND TIMES OF A RAILWAY VILLAGE

I

VESPER PUBLICATIONS,
Old Bilsthorpe, Newark, Notts, NG22 8SN

ISBN 0 9526171 0 2
British Library CIP Data – A catalogue record for
this book is available from the British Library.

DEDICATION

This book is dedicated to three grand old railwaymen - my late father, Leonard Little (1889 - 1976), Harold Watts (b.1904) and Jack Mallett (b.1905). Happily Harold and Jack are still with us; I hope they enjoy the book.

FOREWORD

Railway Towns appeared quite suddenly on the map of England around the middle of the last century, as the rapidly expanding railway companies realised the need for centralised Works for the erection and repair of steam locomotives.

Such Works were generally located well away from existing centres of population to take advantage of lower land values, and new towns had to be quickly planned and built to meet the housing and social needs of the labour force. In this way, places such as Doncaster, Crewe and Swindon soon became household names, and their development and history have been well documented.

Railway villages had exactly the same *raison d'etre* but were obviously much smaller, catering for the more modest employment needs of a major junction or a new marshalling yard. Woodford is perhaps the best known, but there were many others, mostly destined to grow, prosper for a while and then fade away, virtually unknown and unmourned except in their immediate locality.

Langwith Junction was just such a place. Its lifespan as a railway village was less than eighty years, hardly more than that of an average man; being remote from any main line it received little recognition in railway literature. Yet during those eighty-odd years, hundreds of men spent their whole working lives there, unremarked and with scant reward.

This book is their modest memorial.

CONTENTS

1. Introduction 4
2. Passenger Services 9
3. The School Train 10
4. Local Freight 15
5. Coal Traffic 15
6. Trip Working 16
7. Diversions 17
8. Water 18
9. "The Loco" 19
10. Loco Residents 25
11. Visitors 29
12. Sentinels (and others) 31
13. Life at the Loco 31
14. Links and Lodging 36
15. Life on the Footplate (I) 38
16. Life on the Footplate (II) 44
17. After Nationalisation 44
18. Bricks, Mortar and Steel 50
19. The Village 55
20. The Mission 57
21. Personalities 58
22. The Wagon Works 61
23. Life on the Track (I) 64
24. Life on the Track (II) 66
25. The Final Years 66
26. Postscript 73

Appendix 1 Langwith Shed Allocations 74

Appendix 2 Booked Departures ex Langwith Loco. Winter 1946 76

Appendix 3 Booked Departures ex Langwith Loco. Winter 1962 78

Appendix 4 Passenger Train Departures, Shirebrook North 80

Appendix 5 Rainy Day at Langwith Junction 82

Appendix 6 Langwith Loco - The Final Year 84

1. INTRODUCTION

Even forty years ago, when the railway network of this country was still much the same as in pre-Grouping days, most railway enthusiasts would have been hard-pressed to locate Langwith Junction on a map. It wasn't on a main route to anywhere, no famous expresses ever visited its platforms, even named locomotives were a distinct rarity.

Yet it was a fascinating place for all that; I lived within a hundred yards of the tracks from 1934 to the mid-fifties, and my father spent the whole of his working life there, starting out with the Great Central in 1919 and almost seeing out steam on British Rail. Apart from hanging around the station, I also travelled to school daily by train from 1945 to 1951, and this is the period I remember most clearly.

Langwith Junction, nowadays a substantial village, was a green field site on the Nottinghamshire/ Derbyshire border when the Lancashire Derbyshire & East Coast Railway chose it as an appropriate place to throw off a branch line northwards to Beighton, on the outskirts of Sheffield. That was in 1897; ten years later the little line was taken over by the Great Central, and subsequently of course by the LNER. *That* august concern decided that "Langwith Junction" wasn't a sufficiently dignified name for a passenger station, and promptly renamed it "Shirebrook North (for Langwith)", though it was a long mile to the town of Shirebrook, and even further to Langwith! Curiously, the signal box retained its board lettered "Langwith Jc."(not Jct.) and the nearby Loco Depot was always "Langwith"! At any rate, it was always "The Junction" to the locals.

(2) *A general view of the Junction looking west, after a visit from the painters! The crowds visible on Platform 2, and the corridor stock in Platform 3, confirm the arrival of an R.C.T.S. rail tour; the date is 24th July 1960. The spare coach is parked in the disused cattle dock; the 'dolly' (ground signal) Number 46 in front of it had once been one of the most-used at the Junction, as it controlled the reversal of light engines from Platform 1 into the Loco Yard. The original lattice footbridge will be replaced later in the year.*

RAILWAYS OF NORTH NOTTINGHAMSHIRE
WORKSOP
BEIGHTON
SHEFFIELD
LDECR
DUCKMANTON JUNCTIONS
MARKHAM JCT.
CHESTERFIELD MARKET PLACE
LANGWITH JUNCTION
LANGWITH
TO LINCOLN
LDECR
NORTH
ARKWRIGHT TOWN
BOLSOVER
SCARCLIFFE
SHIREBROOK
WEST
WARSOP
EDWINSTOWE
SOUTH
GCR
GNR
MR
MANSFIELD
PLEASLEY
NOTTINGHAM VICTORIA
SUTTON & NOTTINGHAM VICTORIA
MANSFIELD & NOTTINGHAM MIDLAND
LNER
LMS
COLLIERY
RDAJ 10/94

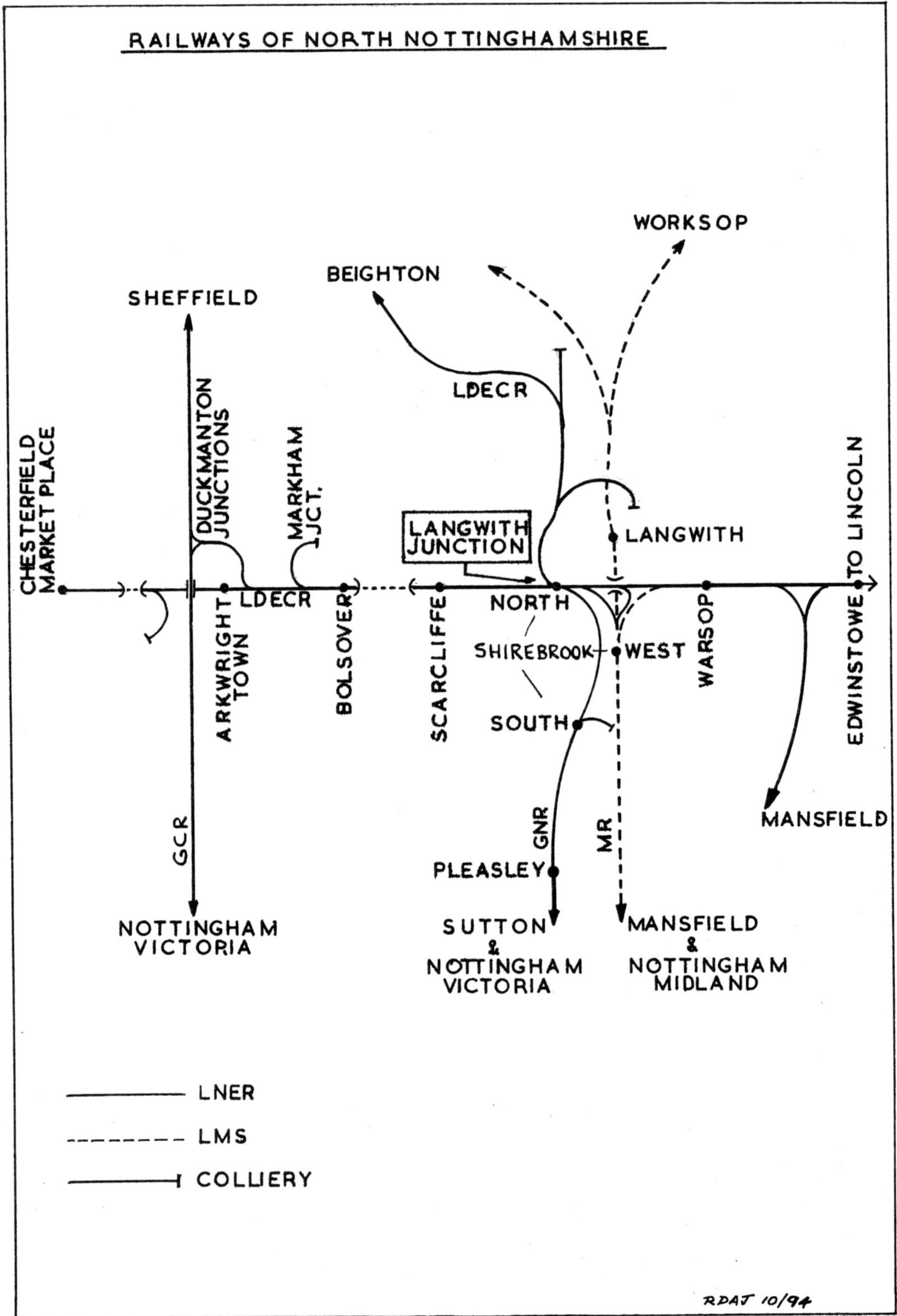

DIA. 1

dshaw Wood
Disused Quarry
Disused Limekilns
Pond
Shirebrook North Station
ROAD
Station House
Goods Shed
Crane
Allotment Gardens
Mission Hall
East View
P O
T C B
B M 347·17
BURLINGTON AVENUE
B M 345·96
Junction
Brookfield County Junior Mixed School
Recreation Ground
Playing Field
Playing Field
Bowling Green
Pavilion
STUBBINS WOOD LANE
RH
18ft R H

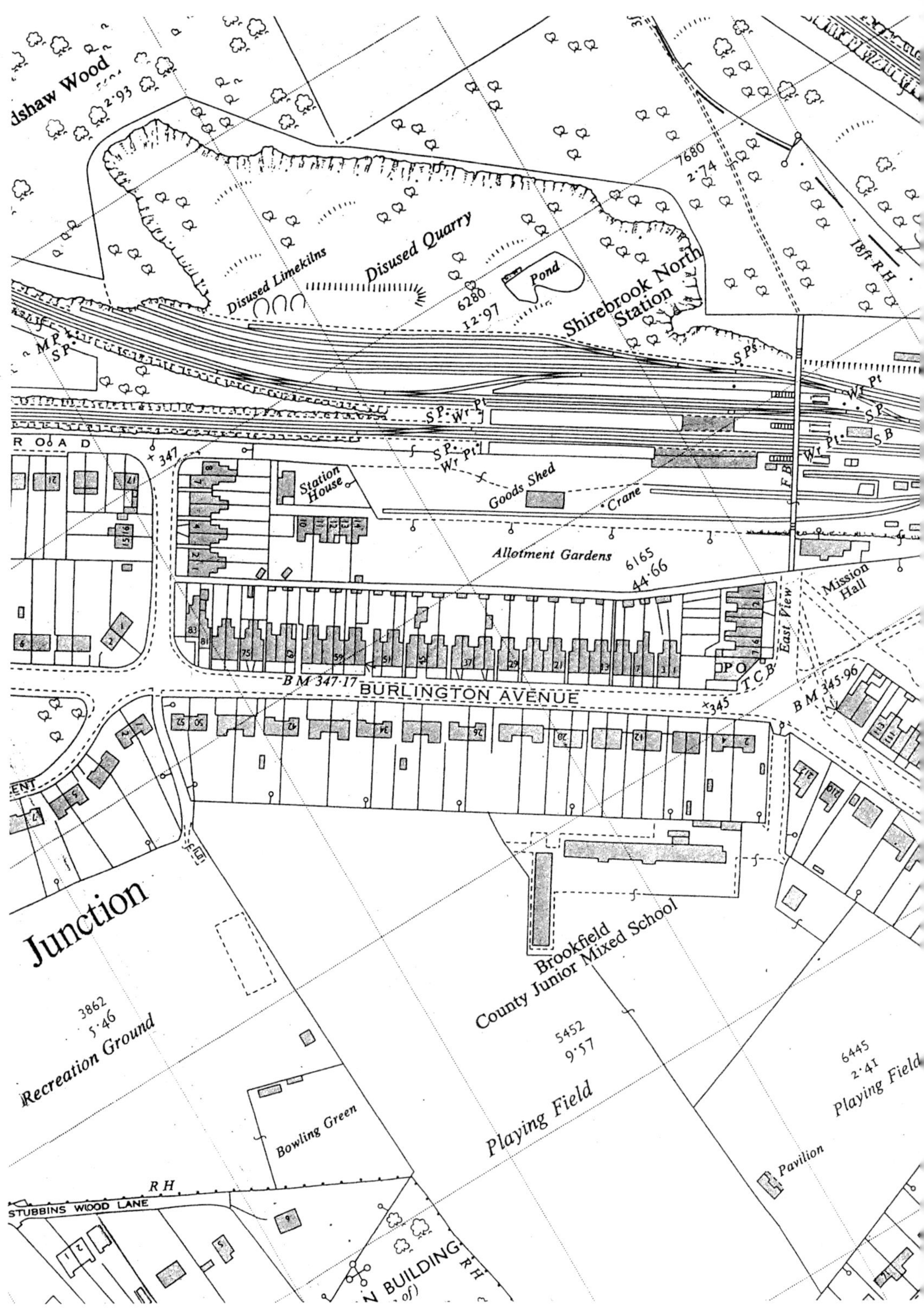

DIA. 2

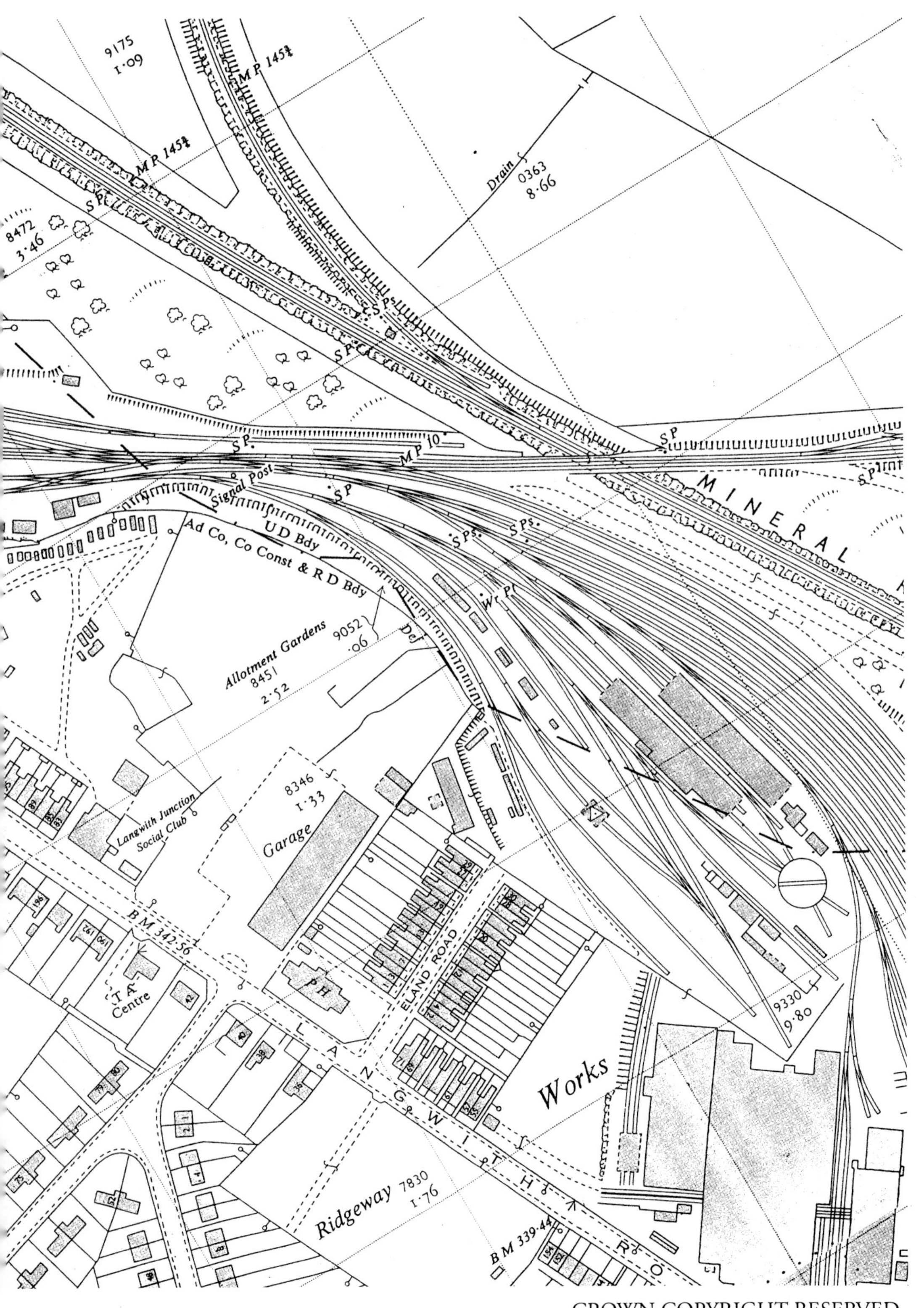

CROWN COPYRIGHT RESERVED

(3) *A '60's view of the Junction looking east; Lincoln line in centre, Beighton Branch coming in behind signal-box, Leen Valley line swinging away to the right in distance. The curious structure on the extreme left is an S&T store; the Permanent Way hut and trolley shed are just visible behind bracket signal in centre.*

As a boy I spent many hours on the station; a friend and I would sit under the canopy on Platform 1 studying the contents of our Ian Allan ABC's - with the retentive memory of twelve-year-olds we could effortlessly recite the tractive effort and dimensions of every class in the book, or identify any locomotive by its number. However, the platform location wasn't actually the best place to see all the action, so on sunny days (there seemed to be more of them then) we could be found perched on the ash-and-bramble bankside next to the junction points; from here we could observe not only all the passing trains but also the continuous movement of engines in the shed yard. Sometimes we would walk down to "The Loco" as the engine sheds were universally called - it was a dirty, inhospitable place, not that that bothered us, but our presence there was definitely unwelcome. The Duty Foreman would soon emerge from his office with arms waving, bellowing for us to get back behind the fence! So Loco visits tended to be left until Sunday afternoons, when we could slip quietly along the rows of slumbering engines grabbing more numbers for our notebooks.

Sunday was the only day of rest at the Junction; for the other six days it was a constant hive of activity - indeed, one of my clearest memories is just *how* busy it was; there would rarely be a moment when something wasn't moving somewhere, if not a train passing, then a pilot shunting the yard or a string of light engines coming on or off shed. All these movements, apart from a few down at the far end of the Loco Yard, would require the attention of the signalman. I knew Derek Talbot, one of the regulars, and it is typical of the skill and knowledge of such men that, years after the box was no more, Derek could sit down and draw out in perfect detail the whole complex track layout, along with every point and signal correctly numbered, and even a sketch of the box interior showing all the instruments!

2. PASSENGER SERVICES

(4) *A c1950 shot of J11 0-6-0 64321, a Langwith resident from before 1932 to 1954, standing in Platform 2 at Chesterfield (Market Place) Station on an unusually lengthy train for Lincoln; two older coaches have been added to the almost new standard Thompson suburban set, suggesting that the occasion is a Bank Holiday, or perhaps the final day of service; the absence of surplus coaches stored in the nearer platform tends to confirm the latter.*

The traffic which Derek and his colleagues controlled was heavy, but not particularly diverse. On the passenger side, the glory days were long gone, though local people still recalled with some pride the years past when, several times each day, four trains would simultaneously occupy a platform, bound respectively for Chesterfield, Lincoln, Sheffield and Mansfield. Services to the two latter destinations had ceased in the 1930's, and all that remained by my time was the sparse Chesterfield - Lincoln run, maintained in its final years by one of the Robinson A5 4-6-2T's and a selection of generally nondescript coaches, most of them clerestory-roofed and a few still gas-lit! (A gas-tank wagon lived at the buffer stops in the centre release road at Chesterfield Market Place to replenish the supply on these coaches). A rare example of modernisation came to this LNER backwater around 1948 with the arrival of a new three-coach set of Thompson "suburbans" - two brake-thirds and a semi-corridor composite; these had steel bodies but were still finished in the traditional teak, painted on complete with graining and knotholes! I used these coaches occasionally on trips to Edwinstowe to see my grandparents, and still recall marvelling at the opulence of the plush upholstery and individual reading lamps - a far cry from the usual ancient horsehair-stuffed relics with their leaky clerestories and big ceiling-mounted glass globe lights, the latter usually half filled with water and dead flies!

There were a few other passenger trains to be seen at the Junction, in the form of excursions either to the East Coast or for football supporters or shoppers wishing to visit Nottingham or Sheffield. On summer Saturdays the best of Langwith's J11's would take a day off from hauling coal and cheerfully set off with quite lengthy strings of assorted stock for a day at the seaside or the match - if you were lucky the coaches would be surplus main-line 61' corridor stock, but at times of great pressure Kings Cross would send down some sets of suburban quad-arts; heaven help you if you were incarcerated in one of *those* for two or three hours - more if trains were backed up six deep on the Skegness Branch!

3. THE SCHOOL TRAIN

My main railway travels, however, were in the period 1945 to 1951, when I attended school each day at Chesterfield, entailing a twenty-mile round trip on the local passenger service from the Junction. There were sixty of us schoolboys, and rather than let us loose among the ordinary passengers, we were provided with a pair of special reserved coaches.

(5) *There are no known photographs of the School Train so here is the next best thing - a model! Nearest to the camera is the GCR saloon, next the NE/NB/LNER hybrid saloon, followed by the Thompson 3-coach set (Brake-Third/ Semi-Corridor Composite/Brake Third). Built by Fred Newman for the author.*

Both were somewhat unusual, being third-class open saloons with longditudinal seating. The older of the two was a flat-roofed vehicle of Great Central origin, one of a batch of three built at Gorton in 1900 for the London Extension. It had a 49' body and featured two long saloons with a dividing partition; at the outer end of each saloon, a pair of cross-seats flanked a central door giving access to a short corridor with toilet and luggage compartments off. The Diagram Reference was 5F1, and the three saloons, which were painted a medium muddy-brown shade, were numbered 51315-17; which one was ours I no longer recall, but when it was withdrawn for repairs following some unusually destructive juvenile escapade, an identical twin was produced the next day.

The other saloon was even more unusual, being built by the LNER to a North Eastern body design with a North British Railway interior! One of a pair built at York in 1925 to Diagram 151B, it sported the classic bow-end roof profile and was finished in varnished teak, genuine this time! The interior comprised three saloons, again with longditudinal seats; the middle saloon still retained its original central table, about eight feet long and with fold-down leaves either side - it proved ideal for halfpenny football!

There were no cross-seats in this vehicle; a centre door in one end division gave access to a cross-vestibule, whilst at the other end, one side seat was shortened to allow access to a door into a short side corridor with again, luggage and toilet compartments off. The two coaches were numbered 3165/66; again which one we had I have not been able to establish.

The two saloons were attached to the front of the 8.27 a.m. local passenger from the Junction, with arrival at 8.56 a.m. giving us little time to dash across town for school opening at 9 o'clock. Similarly in the afternoon there was another rush to catch the 4 p.m. departure; this time the saloons would be on the rear, having spent the day parked at the bufferstops on Platform One.

This parking practice once led to what could have been a very serious accident. There was the usual release crossover near the buffers, and on this occasion the crossover points had somehow become reversed. When the train pulled out, the rear coach (the NER hybrid saloon) followed the crossover onto the adjacent goods road, promptly pulling the other saloon off the track! To make matters worse, the footplate crew neglected to follow the standard procedure of looking back to check the train out of the station.

Fortunately the couplings held firm and the NER coach continued to run along the goods road holding its companion more or less upright, but at an angle of about thirty degrees to the platform. As a School Prefect I had to sit in one of the cross-seats to keep an eye on the younger boys, and I still quite clearly recall hanging onto the seat arm watching the dust rising from the floor in distinct puffs as we bounced heavily from sleeper to sleeper. Through the large plate-glass side windows I could see the barking A5 and its oblivious crew as the platform curved sharply to the left - unfortunately the crew didnt see *us*!

Finally fate took a hand; the outer section of the platform was edged with wood rather than the usual stone flags, and when the derailed coach reached this part, the nearside leading buffer dug into the wood and brought the train to a halt. If the coach had turned over the windows would inevitably have smashed and I wouldn't have given much for our chances; as it was we were able to scramble out just shaken and stirred - that was *one* evening they let us travel in the ordinary coaches!

The morning train was quite often cancelled without warning, due to engineering problems either in Bolsover Tunnel or with a troublesome viaduct over an LMS branch near Bolsover - Midland crews were said to refuse to work underneath if there was an LNER train passing overhead, due to the danger from falling masonry! Such cancellations resulted in passengers being crammed into a couple of East Midland buses for the journey to Chesterfield. Once only one vehicle turned up, an ancient AEC coach which had been converted during the war into a 'crush-loader', i.e. with perimeter seats all round the sides. All sixty of us plus a handful of ordinary passengers were pushed into the unfortunate bus, which staggered off in clouds of smoke; we missed Morning Assembly that day! Another time an old Guy double-decker arrived; unlike most East Midlands it was a 'high-bridge' model and the driver was obviously nervous of the extra height as we approached the restricted clearance under the Midland main line at Chesterfield; we crawled slowly underneath and midway through there was a loud rumble! The driver did a crash stop and jumped out, to find it was a train passing overhead!

Another day a different crew arrived with a Leyland 'decker. For some reason the driver took a different route through Chesterfield and, getting too close to the kerb on Low Pavement, took out all the upstairs nearside windows on an overhanging sign! Thoroughly demoralised, he finished the job by trying to park under the single-deck clearance of the canopy outside the railway station!

(6) *Undated view of Chesterfield (Market Place) Station with J11-hauled up passenger about to depart from platform 4. There are 'large-lettered' wagons in the yard; the Town Hall in the background (built 1936) looks fresh so the date is probably 1937-38.*

If we were 'kept in' at school, there was another train at 6.15 p.m.; an evening function or a rare visit to the cinema saw us catching the last train at 9.00 p.m. which we called "the express" because it omitted the usual stops at Arkwright Town and Scarcliffe (though it was still allowed twenty-six minutes for the ten-mile journey). One evening I was alone on this train and became seriously alarmed when the driver showed no sign of slowing down for the passage of Bolsover Tunnel. The tunnel was slowly being crushed by subsidence and clearances were becoming very restricted, but on this occasion the crew must have been ready for their supper and we rattled smartly through the smoky bore; I was scared enough to move away from the side of the carriage nearest the tunnel wall! Not long afterwards a train actually struck the wall, and the down line was then slewed into the centre of the bore; the up line remained in place but could only be used by platelayers trolleys as the usual 'six-foot' had been halved.

For some reason drama never seemed to be far away on this short journey; one morning we came upon a mechanical digger which was working to clear a landslip in the cutting west of Markham Junction. It was well clear of the train, but as we passed, the vibration apparently caused the digger tracks to slip, and its driver was thrown out, striking his head on the side of our engine. We made an emergency stop so that the unfortunate man could be placed in the Brake Compartment and rushed to Chesterfield, but sadly he later died.

Happily most incidents were less traumatic, like the occasion when our homeward-bound J11 lost both injectors in Bolsover Tunnel and had to retire to Scarcliffe Goods Yard to drop its fire; we had some rare haulage behind an O4 that night!

The Market Place station at Chesterfield was always far too large for the modest traffic levels, and Platforms 4 and 5 were permanently occupied by stored surplus coaches, mostly old GCR clerestories but including also a 'Barnum' or two. Well, not quite permanently - for a few days in 1948 they were all cleared away somewhere, so that the platforms could be used for passenger trains, the normal platforms (1 and 2, there was no Platform 3) being occupied by a display of rolling-stock in connection with the Stephenson Centenary celebrations. The display included one of the Coronation 'beaver-tail' observation cars and several full brakes with displays of models of LNER equipment - I remember being fascinated by a large-scale model of a roll-on train ferry. The adjacent Goods Yard hosted a display of steam locomotives, including the LMS Rebuilt Patriot 4-6-0 45529 which was named 'Stephenson' to mark the occasion. The ancient 1838 'Lion' of Titfield Thunderbolt fame was also present, along with a couple of Midland Railway relics from Derby, but as LNER enthusiasts *we* were more interested in the 'home team', the freshly-outshopped Great Central 'Director' "Prince George" and B1 4-6-0 61085, both resplendent in lined-out black livery.

My daily journeys to Chesterfield ended in July 1951, and before the end of that year the railway had abandoned the unequal struggle to keep Bolsover Tunnel open; the section from Bolsover to Langwith Junction closed completely in December 1951, though the line from Chesterfield to Duckmanton Junction remained open for the still lucrative goods traffic until March 1957.

It was, though we didn't know it then, the beginning of the end for the Junction.

(7) *The Stephenson Centenary exhibition, Chesterfield (Market Place) Station Goods Yard August 1948. Visible are B1 4-6-0 61085, D10 4-4-0 62658 "Prince George", 45529 "Stephenson", and MR 2-4-0 158A.*

4. LOCAL FREIGHT

On the goods side, a pick-up freight left the Junction at 5.50 a.m. daily for Chesterfield, where the loco shunted the yard before returning at 10.30 a.m. Traffic was mainly agricultural produce, or finished timber for Arnold Laver's large yard at West Bars next to the Market Place station; there was also a jam factory with its own private siding at Bolsover.

Of course neither the local freight nor the passenger receipts justified the extensive facilities at the Junction - it was *coal* which paid the bills!

5. COAL TRAFFIC

When first built, Langwith Junction was situated on the eastern edge of the Nottinghamshire coalfield, where it was ideally placed to channel the expanding output away on the first stage of its journey to distant markets. The Great Eastern Railway was a major shareholder in the infant LD&ECR, and therefore much of the coal headed east to Lincoln thence south via March to London; GER locos worked through on many trains, and one batch of 0-6-0's was even specially fitted with additional sandboxes to help cope with the LD&EC gradients.

After 1907, when the local line was absorbed by the Great Central, new junctions were put in at Duckmanton, just east of Arkwright Town, to allow access to the GCR main line; even earlier, the Great Northern had extended its Leen Valley line northwards to Langwith Junction, and connections were also made with the nearby Midland Line at Shirebrook, giving still further outlets for the coal traffic.

A later development was the generally eastward spread of the coalfield as new technology allowed deeper mines to be sunk, culminating as recently as 1965 with the Bevercotes mine near Ollerton. This eastward trend eventually saw Langwith Junction being in the centre of the coalfield, with both loaded and empty trains going off in all directions.

Of course, not *all* coal traffic actually passed through the Junction. There was a constant movement of light engines going off shed to pick up loaded trains at Thoresby or Ollerton to the east, or Shirebrook and Pleasley to the south, none of which were routed through the Junction. At the end of their shifts the light engines would reappear, often coupled in threes or fours with a single brake van for the Guards, and come clanking smartly into Platform 1; from there they would set back into the Down Yard where the van would be dropped off before the engines disappeared into the Loco for disposal. When the Yard began to be cluttered with vans, one of the N5 pilots would round them up and place them in the 'Van Road' behind the cattle dock, where they were convenient for collection by outgoing engines.

6. TRIP WORKING

(8) *Class O4/3 63842, recently overhauled, brings a coal train off the Beighton Branch, 10th March 1961. By this date few O4's were surviving a visit to Works, but this one had been lucky; still in her original 1919-built condition, she would soldier on at Langwith until April 1965.*

The main flow of loaded coal wagons was from the Beighton Branch, originating at Langwith or Cresswell Collieries, and worked by local trip engines, usually an O4 but sometimes a J11 or Q1, into the Up Yard at the Junction; this was generally known as "Newtons Sidings", from the name of the long-abandoned limestone quarry alongside. Reference to the track diagram will show how, by using a facing crossover just to the west of the station platforms, trains could enter the Up Yard without reversing, and run straight up the headshunt prior to setting back into the required road. Trip engines would normally leave their trains there and go straight off for a set of empties from either the Down yard behind the Loco Sheds, or from Warsop Sidings just to the east.

A somewhat unusual feature of Newton's Sidings was that some of the tracks terminated up against the rock face at the end of the quarry; to avoid the expense of bufferstops the rock had been levelled off at buffer height to form an immovable stopblock (the same arrangement had been applied at the end of the Goods Shed road on the Down side). It was standard practice to leave the Brake Vans on trains in Newton's, so there would usually be a line of vans up against the rock face. Often the stoves would still be lit, and as the site was well away from the station, the vans would provide an ideal "hide-away" for my friends and I, though in those far-off innocent days it would be for nothing more than a make-believe journey or as shelter from the rain.

The most northerly track in Newton's had once served a limestone loading gantry and a couple of limekilns, and was shorter than the other roads, hence it was rarely used apart from the odd "cripple"; another track used for the same purpose trailed off the headshunt to terminate in a heap of old ballast under the footbridge (originally it had extended right into the quarry).

Westbound trains leaving Warsop Sidings faced a fierce upgrade of around 1 in 70 and so usually needed a banker; one of the N5 pilots would loose-buffer up to the Brake and push the train as far as the station platforms where the gradient eased, before dropping off and returning to Warsop via the centre crossover. If the Lincoln-Chesterfield Down line was occupied at the time, trains would work wrong-line out of Warsop Sidings and use the same crossover to gain the correct track on the Beighton Branch.

When there was a suitable gap in the traffic, an N5 would come across to Newton's and sort the full wagons before moving them down to Warsop for despatch eastwards. If the wagons were routed along the Leen Valley line a train engine would take them straight out of the yard and away up the bank past Shirebrook.

Loaded trains off the Chesterfield line, originating at Markham, Bolsover or Arkwright Collieries, would work straight through the Junction into Warsop Sidings (always referred to as "The Loops", as they were double-ended). A full train coming down the Leen Valley line and also destined for the Loops would require a more complicated manoeuvre. First the train would run onto the Chesterfield line until the rear end was opposite the signal box; then a Pilot would come onto the Brake Van and remove it out of the way, usually into Platform 2 if there was no other traffic about. Then the Pilot would return to the rear of the train, couple on, and with much crowing of whistles the two engines would ease the train down into the Loops. Once it had come to a stand on the 1-in-70, the train engine would be detached before retrieving its Brake and going off for another load.

Trains coming down the Leen Valley line had to negotiate the long falling gradient through Skegby and Shirebrook South before reaching the Junction and it was not unknown for unfitted trains to have problems keeping under control. On one occasion a freight headed by an ex-LD M1 0-6-4T ran away and hurtled through the Junction and well up the bank towards Scarcliffe before being stopped! At a later date an out-of-control J39 0-6-0 was less fortunate; it reached the Junction just as the stock for the 8.27 a.m. Passenger was coming out of the Down Sidings - the C12 escaped but the leading coach was tipped off the track by the runaway "Standard".

7. DIVERSIONS

A welcome break in the procession of coal trains occurred when diversions were running. The Beighton Branch connected with the Great Central main line at Killamarsh Junction, near Sheffield, and the Leen Valley line did the same at Kirkby South Junction near Mansfield, so the two together formed a very useful continuous loop off the main line, which was regularly used both to relieve congestion and also to avoid Engineers Possessions. The scene would repeat many times - first the Up Home on the Beighton line would come off; then after a suitable pause the sound of a labouring locomotive would be heard approaching through the deep cutting from the north. Eventually it would stagger into view, smokebox often glowing red with the effort, and slowly pick up speed as the gradient eased through the station. The engine, usually a Q6 or a B16 (types not otherwise seen at the Junction), would not be eased, as the driver would be aware of the long bank up past Shirebrook South ahead of him; the exhaust would rise to a crescendo as the train accelerated through the junction points before disappearing behind the Loco. Briefly muffled, the sounds would strengthen again as the train tackled the bank before fading into the distance.

(9) *Class O4/3 63754, then allocated to Hull (Dairycoates), makes light work of Shirebrook Bank with a short Annesley-bound freight on 10th June 1950. Subsequently rebuilt with round-topped boiler to Class O4/8, the old ROD would survive until February 1964.*

Occasionally there would be a little variety; sometimes the approaching exhaust would be heavily syncophated, heralding the appearance of a V2 2-6-2; more often, the train would slow down as it approached the station, prior to stopping in Platform Four for water - this was an unwelcome move because Langwith water was (and is) notoriously hard, but more immediately important to the driver was the prospect of a slow start at Shirebrook Bank!

Diversion traffic often consisted of full trains of flat wagons carrying steel plate from South Wales to the shipyards of the North East, or returning empties. On Sundays even passenger trains would appear, their occupants looking understandably puzzled when they saw Shirebrook North when they should have been approaching Nottingham Victoria! Such trains would usually merit a B1 or V2, but if we were lucky an A3 might appear - once it was "Flying Scotsman".

8. WATER

Mention of Langwith water reminds me that the main supply for the Loco and station was obtained from a small railway-owned reservoir situated on the south side of the line just east of Scarcliffe station;

it was fed by the infant River Poulter which flowed nearby. One of Dad's jobs as a Lengthman was the annual emptying of the reservoir to clean out the filters, carried out at a weekend when the disruption to water supplies would cause minimum inconvenience. A perk of the job was that Dad and his mates could keep all the fish they found (though they always put a few back to ensure a supply the following year!).

(10) *Class K2 61763 takes water in Platform 4 whilst heading an Up coal train. (13th June 1950)*

The water was fed through a 3" main from the reservoir to a 35,000-gallon steel tank, mounted on a brick base on the Down side opposite the signal box at the Junction. There it was mixed with mains water (which was somewhat softer) before being fed to the various water columns - four at the station and three in the Loco. Most of the columns were a plain inverted 'U' design, but the one on Platform Four was an extended swan-neck pattern so that it could be used by engines standing on the back road into the Up Yard as well as those in the Platform.

To minimise the use of mains water, the ever cost-conscious LNER arranged that only Scarcliffe water was used for boiler washing. The hard water must have caused endless problems for the local boilersmiths but the LNER never had the inclination (or perhaps sufficient funds) to do much about the problem and it was left to BR to introduce a water-softening plant, fitted in the brick base of the water tank.

9. "THE LOCO"

When the LD&ECR was first built, the main Works and loco sheds were located at Tuxford, where the line crossed the Great Northern East Coast Main Line; Langwith only merited a small shed for the two shunting tanks allocated to the local yards. As coal traffic developed, Langwith grew in importance and eventually became the hub of the system; although the Works remained at Tuxford until closure in

SKETCH PLAN : LANGWITH LOCO. DEPOT
Down Yard
(Seven tracks)
To
Wagon
Works
Key :
A Engine Shed
B Old Carriage Shed
C Boiler House
D Inspection Pits
E Sheerlegs
F Ash Pit
G Sand House
H Old Coach Bodies
J Foremans Office
K Offices
L Coaling Plant
M Coal stocking Ground
N Ash Wagon Road
W Water Columns
X Site of old turntable

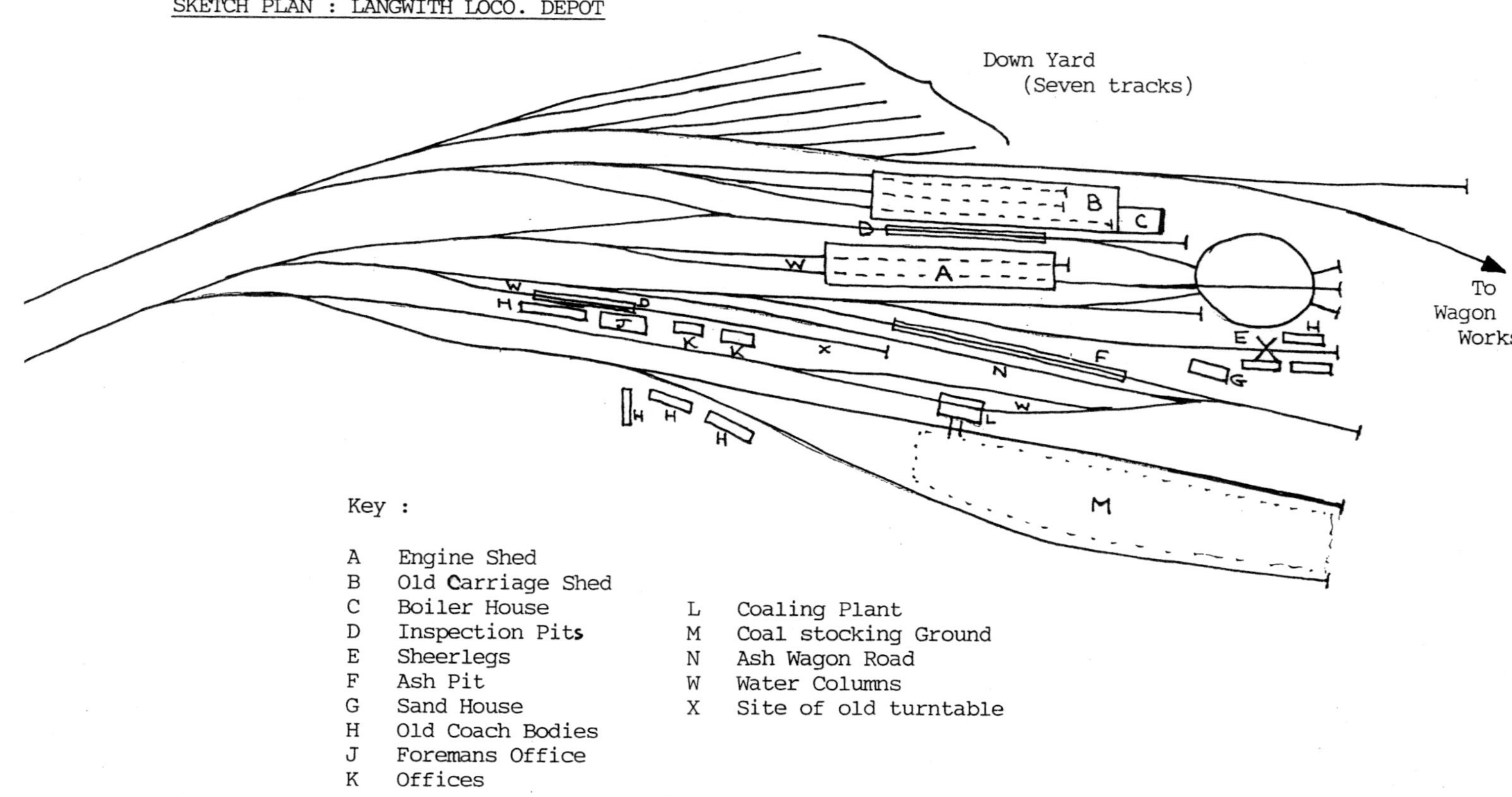

DIA. 3

LNER days, most of the locomotives were transferred to Langwith, where larger premises were soon required. The original small shed was demolished and replaced by a larger one a little to the east, providing covered accommodation for up to fourteen tank locos. Later, after takeover by the Great Central, tender engines began to appear at the Junction, and the new shed soon became overcrowded. The engine shed was never extended, but the wooden Carriage Shed alongside was partly taken over for repairs; it covered three tracks but was ill-suited to its new role, the sides of the structure being open from ground level up to a height of about four feet (presumably to allow ventilation when used for coach storage); this feature was never altered, and working conditions in winter must have been horrendous!

In 1935, thanks to a low-interest Loan from the Government, the LNER was able to proceed with some long-overdue capital works, one of which was the erection of a new coaling plant at the Junction. The towering concrete structure, built by the Mitchell Conveyor and Transport Company of Atlantic House, London EC1, was like others of its type always known as "The Cenotaph", due to a similarity with the shape of the National War Memorial in London. It was filled by hoisting a wagon up the side and tipping it bodily into a hopper at the top, the process producing a characteristic roar which could be heard for miles!

(11) *A good view of the Coaling Plant and the south side of the Loco Depot on 7th March 1964, with O4/7 63843 (destined to be one of Langwiths last 04's) prominent in the foreground.*

A further belated improvement occurred in 1945 when the original turntable was replaced. It was a small affair which had been lengthened several times over the years by the simple expedient of extending the rails over the edge of the pit, an unsatisfactory arrangement which perhaps explains why the shed retained for many years a stud of Q4 0-8-0's which were shorter in the wheelbase than the more powerful O4 2-8-0's. A solution was eventually found when the shed at Sheffield (Neepsend) closed in 1944; the redundant turntable from there was sent to Doncaster for refurbishment before being delivered to Langwith. One Sunday morning the Colwick crane was sent over to help the local one lift the table into place, a new pit having been dug to the east of the shed building where the turntable could serve three tracks rather than one as previously.

Apart from this, there were few signs of modernisation at Langwith. The LNER was obviously satisfied that such facilities as offices and workshop accommodation could quite easily be met by the allocation of surplus old coaches, the grounded bodies of which could be found all over the Langwith depot site. Only one could be positively identified - LNER 4543, a Great Northern 63' clerestory brake-third, built in 1904, found its last resting place at the west end of the shed yard in 1950 to serve as an enginemen's locker room, still in teak livery with shaded numbers and lettering. There were other bodies near the Eland Road entrance which were used for First Aid and Mutual Improvement classes, and several more, even older, grouped round the sheerlegs to form primitive workshops.

(12) *A 1960's view of the sheerlegs and Coaling Plant; note the ancient coach body used as a workshop, also the new extension to the Wagon Works in the left background.*

The local crane has already been mentioned. This formed part of the Depot Breakdown Train, which resided when not in use on the most southerly track in the old Carriage Shed. When I last checked, in 1962, the crane was an ancient 15-tonner built by Cowans Sheldon as long ago as 1898, supported by an equally venerable 49' clerestory coach, a GNR 6-wheel flat-roofed Passenger Brake, and a Great Central matchboarded coach lettered "Staff Riding Van". Soon afterwards the crane was replaced by a heavier 25-tonner previously at Middlesbrough, and later still when this crane went to York along with the 6-Wheel Brake for preservation, a 30-ton long-jib replacement arrived; this only lasted a couple of days, and the staff were still weighing up the new arrival when word arrived that it was needed elsewhere. Thereafter the Diesel Depot, as it had become by then, had to manage with Kelbus jacks when attending a derailment.

(13) *A rather gloomy shot, but the only one known of the Langwith Breakdown Train going out on a call. 9F 92146 (note the express headlamp code) hurries off with the train formation as described in the text. 22nd May 1965.*

(14) *The Great Northern 6-wheel Brake which owes its survival to being used in the Langwith Breakdown Train; seen at Shirebrook Diesel Depot in 1970, it was later claimed for the National Collection and moved to York for preservation.*

Near the coaling plant an area of open ground was set aside as a coal-stocking yard, the railways still following the practice of buying coal at cheap summer rates and storing it for use the following winter. The arrangement was to build up the larger lumps in dry-stone-wall fashion, then throw the smaller stuff in behind, the "walls" being white-washed to discourage pilfering. The whole laborious business was done entirely by hand, and the reverse process repeated later in the year; the cost in terms of scarce labour, must have been more than the saving in price!

Other facilities at the Loco Depot included a small boiler house built on to the end of the Carriage Shed, housing an ancient ex-locomotive boiler derated to produce steam and hot water for boiler washing; a tall chimney had been added to improve the draughting, leading to the inevitable nickname "The Rocket". Outside was a short siding for wagons delivering stores; brake blocks and boiler tubes were racked inside the Carriage Shed, two of the roads being shortened to allow space. Stores wagons could also be parked, along with the snowplough (based on an old GNR tender), on the over-run tracks beyond the turntable.

(15) *Not the Langwith snow-plough but identical in appearance, DE 330915 was based at Kirby (Midland) shed; it's Great Northern origins are apparent.*

Near the latter was a sand-drying shed with its own furnace, and a sheerlegs for heavy lifting. There were three outdoor pits, two for examinations and repairs and the third for fire-dropping - the latter always seemed to be over-flowing!

The overall impression, even to our young eyes, was a place of appalling working conditions, far removed from the present idealised memories of steam days. Even inside the shed it was always cold, dark and draughty, with the floor ankle-deep in old ashes, broken bricks and oily water. Yet this was a time when new steam locomotives were still being built!

10. THE LOCO RESIDENTS

Appendix 1 gives details of the locomotives allocated to Langwith at various times during the LNER and BR period.

Going back to earlier days, the LD&ECR only ever rostered tank engines, of four varieties - 0-4-4T (LNER Class G3) for passenger work, 0-6-0T (J60) for shunting, and 0-6-2T (N6) and 0-6-4T (M1) for freight traffic. At the time of Grouping in 1923 many of the LD locomotives were still at Langwith, supplemented by various GC classes, but in early LNER days the G3's and J60's, which were becoming too small for the heavier traffic, gradually drifted away to less-onerous duties elsewhere. Similarly the big M1 tanks moved to Tuxford (apart from a single example which remained at Langwith until 1939), leaving only the N6's to see out their days on home ground, albeit reduced to yard shunting duties; as they were withdrawn in the 1930's they were replaced by the somewhat similar GCR N5's.

(16) *No, not an N5, but the dimensionally-similar ex-LD&ECR Class N6. Seen here on 6th June 1936 after being fitted with a Gresley 'flowerpot' chimney, 5043 was the last N6 at the Junction when withdrawn in January 1937; note the newly-built Cenotaph in the background behind the locomotive on the old turntable.*

Freight work was in the hands of "Pom-Poms" (Class J11 0-6-0's) and "Tinies", the wry nickname for the first Great Central heavy freight engines (Q4 0-8-0's). The J11's had a very long innings at Langwith, the last two not being withdrawn until September 1962, but the Q4's were gradually replaced by the rather more powerful O4 2-8-0's, always known at Langwith as "RODs", a reference to the Railway Operating Division of World War One for which many were originally built. In later years when most O4's were rebuilt with round-topped boilers, the Belpaire-firebox originals were sometimes called "Flat-tops".

In October 1947 there was an unprecendented upheaval at Langwith when someone in higher authority decided to clear out almost all the O4's and replace them with a rather smaller number of Gresley O2's; these three-cylinder engines were a more modern design than the RODs, and somewhat more powerful, but the fiercely-partisan Robinson enthusiasts at Langwith could be expected to regard the new arrivals with some suspicion. Fortunately the O2's, nicknamed "Straight-Eights" for some long-forgotten reason, proved highly competent once the locals got used to their 9-foot firegrates and the cabs, which despite their commodious appearance were regarded as cramped by the Langwith men. Apart from their higher tractive effort the three-cylinder layout gave them the benefit, obvious even to us youngsters, of a more even torque which allowed smoother starts - one rarely saw an O2 slipping.

(17) *Between October 1946 and June 1950, freight workings at Langwith were monopolised by Gresley 02 2-8-0's; 63976 takes water, 16th March 1950.*

So they were accepted, but as it happens all in vain; by June 1950 Langwith was experiencing an aggravated shortage of fitters, and the O2's were moved out again, RODs returning to replace them. It appears that the latter were thought to need somewhat less in terms of maintenance than the Gresley design, though by the look of the newcomers they were the scrapings of Mexborough's barrel, some being so rough that the unfortunate fitters were probably no better off!

Two of the Langwith RODs, 63627 and 63809, had their moment of fame in 1952 when, along with three others, they were sold to the War Department. After overhaul and the fitting of new boilers, they were shipped out to the Suez Canal Zone in May 1952, and worked there until W.D. operations were terminated in 1955. They were then transferred to the Egyptian Railways, and some may have survived until c1961.

(18) *Two old Great Central stalwarts "brewing up" at the west end of the loco shed. The original painting by Grimsby artist John Willerton was specially commissioned for presentation to Colin Palmer upon his retirement as Area Manager at Immingham in 1994.*

On the passenger side, the G3's had been replaced by a modest collection of 4-4-2T's, mostly C14's but with a few C12's. By 1935 these were all needed elsewhere, and Langwith had to manage for several years using borrowed N5's, often short of steam with their small grates, or J11's, strong enough but unsuitable for tender-first running. During WWII, many passenger types came and went; first a couple of C14's returned, soon to be replaced by various GNR 4-4-0's (D2 and D3) - neither of the latter were liked locally, partly because they were Great Northern and therefore regarded as feeble by definition, and partly because like the J11's they were unsuitable for tender-first running. There *was* a small turntable at Chesterfield but it was a tiny ancient thing designed for the LD&EC tanks and never replaced, so the enginemen had to sacrifice their layover time to run to Duckmanton Junction to turn on the triangular layout there. Eventually some C12's came back - still feeble but at least they could run either way round.

Also during the War, Langwith had to provide power for a heavy passenger special which ran daily between Mansfield and a munitions factory at Ranskill; they really had to scrape the barrel for *that* job! First came a pair of B2's (the original Robinson type, not the Thompson rebuilds) - they lasted a month. Next to arrive was a brace of B4's (three months), then two D10's (six months), then three D11's (two years), finally some GNR C1 Atlantics, which saw out the War.

Earlier, the depot had hosted several groups of wandering 0-6-0's from the North-East. Longest-lasting were several J27's (1935-39), oddest were some J28's from Hull (eight months in 1937) - these domeless ancients from the Hull & Barnsley had screw couplings and steam heat connections, so found a little light work on passenger trains, especially at weekends; probably the least useful were five J21's which arrived from Boston in June 1939 and survived at Langwith until April '42 when they were packed off back to York. They must have been in a parlous state by then - one of the Fradley brothers, who were both boilersmiths at Langwith, once put his hammer right through the firebox side of one! What the local crews, used to the indestructible Pom-Poms, thought to these little 0-6-0's has perhaps fortunately not been recorded.

Perhaps the only other noteworthy residents in LNER days were the "Twins", a pair of the hefty Q1 0-8-0T's which had been created by Edward Thompson by rebuilding elderly Q4 tender engines. Someone once said that the best thing which came out of *that* exercise was the batch of spare tenders! At any rate, their intended use for heavy shunting wasn't needed locally, so the two were put on local trip work, particularly the short run between Langwith Colliery and Newtons Sidings or Warsop Loops. They were also used for experiments with water-softening "briquettes", the most obvious effects being a coating of yellow slime and a horrendous smell! More often than not, however, they were to be found parked together on the short siding near the turntable; perhaps their water capacity was inadequate for any sort of main-line work.

Incidentally, they weren't identical twins, 9928 being a Q1/1 and 9929 a /2, the latter with larger tanks and bunker, though the difference was only a matter of inches. What they *did* have in common was the unique cast-iron LNER "lozenge" designed by Thompson to avoid the need for transfer lettering. With its light blue background it looked quite smart when ex-works, but very quickly became illegible under the usual coating of grime; it never appeared on any other class, although the prototype EM1 electric for the Woodhead line had a painted version applied to the cab doors.

One final note on shed allocations in general; it seems that while some engines stayed at the same shed for a generation or more, others moved around almost annually. Examples of "permanent" residents at Langwith included three of the J11's - 4289, 4378 and 4414 - all of which were there from before 1932 until the mid-Fifties; three of the four regular N5's had also been on the strength since the mid-thirties, as had several O4's until the great clear-out in 1947.

(19) *One of the "twins", Q1 0-8-0T 69928, crosses the bridge over the Midland line with a loaded coal train for Warsop Loops, 2nd September 1955. Just visible in the foreground is the Great Northern Leen Valley line.*

It is generally agreed that no two steam locomotives perform exactly the same, and I can only assume that it was the less satisfactory ones which were moved on at the first opportunity. If this was the case, Langwith must have had more than its fair share of black sheep!

11. VISITORS

In April 1949 I began supplementing my ABC notes with a record of unusual happenings and visitors to the Junction. One of the first entries, on April 17th, noted the presence of V2 60880 on shed, a rare visit from a top-link engine (the only other recorded example was 60876, which went on shed for water on 29th June 1965); 60880 was under repair after failing on a diversion. Other oddities in 1949 included Great Eastern 0-6-0's 64643 (J19) and 64688 (J20), echoes of long-gone days when "Sweedie" engines worked along the LD&EC line with coal trains for the south. On September 18th I recorded my first sighting of a new K1 2-6-0, 62013 of 39A (Gorton).

A minor mystery which has not been cleared up to this day concerns K3 1950 (or 61950, we always ignored the first digit), which was recorded on shed on April 13th "with red lining". This style of painting had been discontinued by the LNER some ten years previously and officially was never used again apart from a batch of new B1's, but it must have been reasonably fresh for me to bother making a special note. Neither was it a reference to the new-style British Railways mixed traffic livery, since only four days later a second note recorded K3 61852 "with red-cream-grey lining".

1950 also saw some unusual visitors. Great Northern engines were always quite rare at Langwith (C12's and visiting K2's excepted) and the hoarse asthmatic chuff of the earlier Ivatt designs would always bring us running. Visitors in this category included J3 64115 from Boston, and J5's 65481/96.

One type never seen at Langwith, before or since, was the GER D16 4-4-0, two of which, 62588/99, stayed overnight on April 20/21 whilst enroute from East Anglia to Trafford Park.

(20) *Great Northern 4-4-0's had been common enough at Langwith during the War years, but by 26th March 1950, when Boston D2 2181 spent the weekend on shed, they were very rare visitors indeed. This was probably the last visit of a D2 to Langwith – they had all gone by 1951.*

(21) *Known as "Standards" years before the British Rail designs appeared, the J39 0-6-0 was not often seen at Langwith; Lincoln's 64736 spends a quiet Sunday on the Carriage Shed road, 2nd April 1950.*

Specials on Whit Sunday, May 28th, produced B1 61281 and K3's 61894 and 61960 in place of the usual J11's; in later years Langwith was allocated examples of both classes for this type of work. In September, N2 69552 was another unusual visitor, standing on shed overnight in a rather sorry state after derailing and going down the bank near Mansfield on a passenger train; she was being towed to Doncaster for repairs.

1951 saw a "namer" - B17 "Melton Hall" was noted running down Shirebrook Bank light engine on April 20th. Apart from sister engine "Bradford" which Lincoln occasionally turned out for the passenger, the only named engines normally seen at the Junction were sundry elderly GCR 4-6-0's long relegated to freight work.

Early in 1953 I left the Junction for National Service, and though I returned briefly in 1955, my notes were sketchy thereafter.

12. SENTINELS (AND OTHERS)

In the 1930's, the LNER experimented with the products of the Sentinel Waggon Works at Shrewsbury, in the form of steam railcars and also small four-wheeled geared shunting locomotives. The former were reckoned to be more economical for lightly-loaded branchline passenger trains, and the latter, not needing a fireman, offered some savings in light service.

Sentinel steam railcars were seen at the Junction between 1929-31, when they were used on the Nottingham (Victoria) - Chesterfield service which ran via the Leen Valley line. "Commerce" is particularly remembered, but neither she nor her sisters would visit the Loco except in case of failure en-route.

A Sentinel Y1 shunter, number not recorded, was the responsibility of Langwith Loco for a period during WWII, though not officially allocated there. It was on loan to the Bryan Donkin Steelworks at Chesterfield while their German i.c.engined shunter was unserviceable - spare parts were rather hard to get at the time! Because the Y1 was too slow to return to Langwith for servicing, a fitter had to be sent over to Chesterfield each week to maintain it on site.

Another unofficial visitor at about the same time was a Great Eastern J67 0-6-0T, which was loaned to Clipstone Colliery near Mansfield for a while; the little tank did manage to get to Langwith at weekends for boiler washout and servicing.

13. LIFE AT "THE LOCO"

At the beginning of this century it was customary for working-class children to leave school at twelve or thirteen and find a job; Harold Watts was no exception. Harolds father was a Goods Guard on the LNWR at Market Harborough, who moved into No.1 Railway Cottages (the plain terrace near the Goods Shed) after obtaining work with the LD&ECR. Harold was born soon after, in 1904, and after completing his schooling he found work at a local bakery. This gave him little scope to exercise his mechanical aptitude, so when an opportunity arose he joined the Great Central as an Apprentice Fitter.

That was in March 1919, but Harold still clearly remembers his first day at work; reporting to the Foreman Fitter, a Mr. Bradshaw, he was placed in the care of one Bernard Walker, who was working in the smokebox of Q4 0-8-0 1077; Harold's smart new overalls were soon suitably "christened"! Thus began fifty years of hard graft which he still recalls with a mixture of pride and frustration - pride in the knowledge that over the years his skill and hard work 'kept the job going', but frustration that his efforts were made immeasurably harder by the lack of equipment and facilities inherent in a depot which was right at the bottom of the LNER pecking order. For Langwith was no glamour shed; there were no prestige jobs to catch the attention of higher authority, just the hard slog of moving mountains of coal (though ironically this humble task was putting far more into the coffers of the Company than the East Coast streamliners!).

In the early days, virtually all repairs had to be done with hand tools, but very gradually a few machines arrived - lathe, drilling machines, grinders; often castoffs from more affluent sheds, they were installed in makeshift workshops in the lean-to alongside the engine shed, or in old grounded carriage bodies scattered around the depot. Powered by shafting and open belts, they would have given a modern Health & Safety Inspector nightmares!

Harolds first weekly wage was 10/9d. (53p.), but he enjoyed the new mechanical challenges and persevered through a six-year Apprenticeship. In 1925 he 'came out of his time' and as so often happened there was no proper job available locally, so he took a temporary post at the Erecting Shop attached to the Depot at Leeds (Ardsley). The two-road building could accommodate eight engines, shared by four gangs of fitters; the repairs were mainly heavy jobs requiring lifting.

Having become used to the simple, sturdy Great Central locomotives, Harold now found himself working on Great Northern engines which were less to his liking. For example, he remembers the problems involved with the replacement of axlebox brasses, a simple job on a Robinson engine (all that was necessary was to drill and tap a couple of 1/2″ holes, then screw in a pair of bolts to provide lifting handles), but much trickier on the Ivatt designs. One particularly awkward task in this category involved the unique booster-fitted GNR Atlantic No. 4419, the trailing truck brasses of which had to be regularly replaced due to the lack of proper springing, there being no space available for conventional helical or laminated springs. Harold suggested drilling out the brasses to the GCR pattern and this was eventually done, but the engine continued to require constant attention even after Darlington had fitted special rubber suspension units to the trailing truck.

As mentioned, the Ardsley job was a temporary one for nine months only, after which Harold had to move to Gorton, where there was a vacancy in the Running Shed. This put him back among Robinson engines, but although he was well-regarded at Gorton and promised early promotion, Harold still hankered for a return to the Junction, and when an opportunity arose in 1927 he was quick to apply. So he returned, and was destined to remain there, apart from short periods relieving elsewhere, for the next forty years.

Space was always at a premium at Langwith; one of the two roads in the Running Shed was reserved for boiler washouts, so when the other was full any additional work had to be carried out either in the open, or in the old Carriage Shed next door, which wasn't much better! The three-road Carriage Shed accommodated the Breakdown Train and one of the branch passenger sets, but the centre road could be spared for loco repair work, albeit in unbelievably primitive conditions - the wind blew through the open sides, there were no end doors, the floor was ashes and there was no lighting! It was in just such spartan surroundings that Harold, at an early stage in his career, managed to tip an O4 into the pit whilst endeavouring to jack up a driving wheel prior to changing a siderod bush!

In the early days there was still time and labour available to keep engines clean as well as mechanically sound. The passenger tanks, originally exLD G3 0-4-4T's and later C14's, had specially-burnished buffers, and one of Harolds first jobs was to remove these from any engine called into Works, to ensure they didn't "go missing" whilst the engine was away; they would be re-fitted upon its return.

For many years the pick of Langwith's stud of J11 0-6-0's would be used on weekend excursions to East Coast resorts, and engines selected for this duty would enjoy a 'shed day' on the Friday so that they could be fettled round and receive an extra polish. In later years of course, there was little enough labour available for repairs, let alone cleaning (though the Langwith passenger tanks were kept reasonably clean right to the end).

One unusual task which Harold remembers was the replacement, around 1940, of the depot stationary boiler. This ancient contraption lived in a small extension to the Carriage Shed, where it was tended by one of the shed labourers in its undemanding task of supplying hot water and steam for boiler washouts. The original boiler was eventually deemed unfit even for this modest duty, so Harold and his mates were allocated a boiler from a recently-withdrawn M1 0-6-4T; this was quite a large design but unsuitable for use elsewhere, and in any case would be de-rated to 75lbs pressure for its new role. It is interesting to speculate whether this boiler came from the same locomotive whose chassis and wheels were used around Gorton Works for many years thereafter to carry a large air tank.

Other out-of-the-ordinary tasks included attention to the Coaling Plant. Prior to the installation of the new Mechanical Plant in 1935 there was an old wooden inclined-ramp coaling stage, situated where the ashpits were in later years. I have not been able to locate any photographs of this structure, but it could accommodate three wagons under a covered structure at the top of a steeply-graded approach ramp. Climbing this obstacle was always a challenge to the old shed pilot, and on more than one occasion a valiant effort would be made only for the wagons to thunder through the shelter and burst through the bufferstop beyond! It was not unknown, either, for the points at the bottom of the ramp to be inadvertently changed while the pilot was gathering its strength further up the yard, resulting in a wagon or two of coal being deposited in the turntable pit!

Once the modern plant was installed, different problems arose. The new concrete hopper was filled by hoisting wagons on rails up the side of the tower before inverting them in a cradle at the top. In the days of wooden wagons, floor planks were not always securely nailed down, and whilst this had no detrimental effect in normal traffic, once the wagon was turned upside down the planks would follow the coal into the hopper! Inevitably they would then jam the loading chutes, and someone had to make a precarious descent into the depths to clear the obstruction.

With the increasing use of steel wagons this particular problem disappeared, though another nuisance was never eradicated - the loss of axlebox oil from inverted wagons; this dripped down the outside face of the hopper and had to be periodically removed by hoisting a couple of cleaners or labourers up in an empty wagon to scrape the deposits away using a tool rather like a garden hoe.

A regular part of Harold's duties in later years involved control of the Depot Breakdown Train. On average this was called out two or three times each week, not to "disasters" of course but usually to rescue an engine or string of wagons which had been put on the ground in some ill-maintained colliery siding. In later years, sadly, there were also cases of deliberate vandalism; he remembers an instance at Bilsthorpe Colliery where all the keys were removed from a length of track allowing the rail to turn over under the weight of a Type 4 and a dozen loaded coal wagons.

At least with a steam locomotive there would be plenty of solid steel under which to place lifting jacks or slings; the same couldn't be said of the replacement diesels! By comparison these were flimsy things, generally leaking fuel or with delicate pipes or wiring to avoid.

World War II didn't make a lot of difference to Harold's work, apart from the nuisance of working under blackout regulations and of course, the extra pressure placed on the grossly-overworked railway system. The local Home Guard used Newton's Quarry for target practice (the writer recalls years afterwards collecting spent cartridge cases there) and some bombs were dropped nearby, usually by lost Luftwaffe pilots looking for Sheffield. One parachute mine destroyed a cabbage patch next to Dad's allotment, leaving scraps of silken cord and sheet draped everywhere; another initially failed to explode, finally going off during Sunday Morning Service at the Mission - shrapnel landed in the nearby Goods Yard and the explosion was heard in the depths of Bolsover Tunnel several miles away.

A wartime visit by the Royal Train is also recalled; headed by a B7 4-6-0, it was parked overnight in the deep cutting just west of Scarcliffe Station, with instructions that, in the event of an air-raid, it was to be shunted smartly into Bolsover Tunnel!

After the war, a welcome sign of a return to normality came in the brief visit to Langwith of the new L1 2-6-4T No. 9000, its bright apple-green livery in marked contrast to the prevalent grimy black. As part of a series of tests to see if the new design could cope with freight working, the L1 worked a trial from Colwick to Shirebrook followed by a trip from Warsop to Broughton Lane (Sheffield), the latter comprising a Class 7 load. Not surprisingly, although the sturdy tank could cope quite easily with hauling loads normally taken by an O4, when it came to *stopping* it was a different matter! With braking only available on six wheels as against fourteen on a tender engine, the test trains regularly over-ran signals; even without the benefit of hindsight such a result must have been fairly obvious!

As steam began to disappear from British Railways, the old Great Central locomotives which had served faithfully at Langwith for three generations were gradually replaced by more modern designs rendered surplus elsewhere. The 9F 2-10-0's, for instance, were welcomed at the Loco for the convenience of their rocking grates and readily accessible fittings, but they still brought problems. The long rigid wheelbase of the "Space-ships" soon gave trouble on colliery lines and sidings, and the 9F's had to be restricted to "mainline" running only. Like the Britannia's which started running through on coal workings from March (a sad come-down after their sparkling exploits on the Great Eastern main line), the water fillers were found to be too high for the local water columns; the engines were too long for the Langwith turntable too. The Brit's were sent to Clipstone to turn on the triangular junction there, but the 9F's ran tender-first much of the time.

(22) *A job for Harold and his crew: Class O4 63585 disgraces herself in Welbeck Colliery Sidings, all wheels off. (Early 1950's).*

(23) *And another! N2 69552 being recovered after going down the bank at Forest Town, near Mansfield whilst hauling the 9.18 a.m. local passenger train from Nottingham to Edwinstowe on 2nd September 1950.*

(24) *A 1960's night shot of Harold Watts and his Breakdown Crew : from left to right - Harold, Ron Lewis (later to become a long-serving Member of Parliament for Carlisle), Ernie Rayson, Pat Dodds and Johnny Jevons.*

As the years passed, Harold was moving steadily up the promotion ladder - Chargehand Fitter, Mechanical Foreman, finally Mechanical Supervisor. He was also in demand for relief duties elsewhere, including spells on the Running side, and he worked at Tuxford and Barrow Hill (Staveley) at various times.

(25) *Only a few months away from retirement, Harold Watts stands with Driver Ted Mayes in front of the experimental 4000-hp diesel-electric "Kestrel" at Shirebrook Diesel Depot, December 1968. The engine was at that time rostered for two trips daily with coal trains between Mansfield Colliery and Whitemoor, but it was too much, too soon, for British Rail, and was eventually sold to Russia where it survived until the early '90's.*

In February 1966 Langwith finally closed to steam, and its duties and staff moved down to the new Diesel Depot at Shirebrook (West); Harold went too, and despite being near to retirement tackled the new technology with enthusiasm. The new purpose-built workshop which replaced the initial temporary premises was a world away from the primitive conditions which had prevailed at Langwith right to the last - there were even *doors* on the building!

Despite the improved working environment, Harold's most nostalgic memories are still of the days when, covered in grease and coal dust, he laboured hard and long to keep the old Great Central engines out on the job.

14. LINKS AND LODGING

(For this section, I was greatly helped by conversations with retired Driver Jack Mallett, who had started work at Langwith in April 1918, even earlier than Harold. Jack became a fireman in 1923 and moved across the footplate in 1940.)

There were three Goods Links at Langwith Junction, known simply as No.1, No.2 and No.3. Each had twelve sets of men and twelve turns, six being lodging and six out-and-back. Because of food rationing, lodging became impracticable during WWII, and the practice ceased at Langwith at the beginning of October 1941; it was not reinstated when peace returned. (See 'Life on the Footplate' for details of the postwar method of working).

Some of the lodging turns included :
Gorton : five turns daily, ranging from 2 a.m. to 7 p.m. starting. Trips started either at Mansfield Concentration Sidings ("The Con") or Warsop Yard ("The Loops"), and were generally double-headed, the pilot coming off at Sheffield before banking the train up to Dunford. Typically a 2-8-0 would be piloted by an 0-6-0, though a pair of 2-8-0's would be rostered if the load warranted. In later years on trains from Mansfield, the pilot would be attached at Warsop, due to a weak bridge near Clipstone. The usual route would be down the Beighton Branch to Killamarsh Junction, but at least one train daily would run via Markham Junction, attaching traffic there before working via the Duckmanton junctions onto the GC Main Line. The 7 p.m. was known as the Manchester Goods and carried general traffic, the others being coal only.

During the 1928-1930 period an overtime reduction exercise saw men on Gorton turns only working as far as Penistone (Barnsley Jct) where they were relieved.

Whitemoor : five or six turns daily, coal traffic only, again originating from Mansfield or Warsop and running via Lincoln, Sleaford and Spalding.

Woodford : one morning and one afternoon train each day, the Up trains ran via the Leen Valley line, but the morning Down working came via the Great Central main line and Duckmanton Junction before dropping off empties at Markham Jct.

Mexborough : One train daily, departing Warsop at 7 a.m.

Hull : Two trains daily, one each from Mansfield and Warsop.

On lodging turns, the engine, usually an O4 but sometimes a J11 if the load was light, would often lay over with the men before working the return trip.

Lodging Facilities : Gorton had a dormitory in what Jack called "an old school". It contained a dining room and a sleeping area, the latter laid out with a long central corridor and four side vestibules, each giving access to four cubicles each side. The cubicle walls were only head high, so with men arriving and leaving at every hour of the day and night, the opportunities for sleep must have been somewhat restricted!

Immingham was Jack's favourite, being kept very clean by a Mrs Powell; an old man called George covered the night shift from 10 p.m.

Whitemoor had "cooks" on each shift who would warm up the men's food. Woodford used two adjacent three-storey houses quite some distance from the railway; they were very cold in winter with only small open fires on the ground floor.

At Hull, Mexborough and Penistone, the men lodged in private houses which were generally much more comfortable than the railway hostels. The same arrangements applied at Langwith Junction, men staying with Mrs Ogden or Mrs Nettleship on Langwith Road (near "Davis's Bridge"). Mrs Mallett also took lodgers at Burlington Avenue.

The minimum lodging allowance was nine hours, though 12-16 hours could be taken according to timetable demands. Apart from sleeping, the time would be passed eating, playing cards or visiting the cinema.

A driver would usually keep the same fireman for twelve months, but Guards would change from day to day; on lodging turns the Guard would work the whole week with the same footplate crew.

15. LIFE ON THE FOOTPLATE (I)

(In the course of researching this book, I visited retired Driver Mick Newham of Langwith Junction. Mick showed me his father's 'Lodging Box', a sturdy metal container with a domed lid supporting a carrying handle and a stamped brass plate 'F.Newham, Colwick'. Footplatemen used these boxes for their personal items needed when working a two-day trip which required lodging away from home; they were sometimes called "Grimsby Tins" as many were made by a small firm based there.

The box contained a rare find - an almost complete set of Frank's diaries, covering the period 1924 - 1968, and listing each days work, with destination, hours on duty, engine number and mates name. These records, kept meticulously for almost half-a-century, form the basis for this chapter).

Frank Francis Newham was born in 1905 and was so keen on a railway career that he joined the Great Northern Railway before he was old enough to legally work shifts! This temporarily barred him from enrolling as a Locomotive Cleaner, so he worked as a lamp-boy until he was old enough to transfer.

After several years as a Cleaner at Colwick, he qualified as a Passed Cleaner (i.e. able to work on Firing duties) on 1st April 1924, and later that month transferred to Doncaster. His very first footplate job there, the lowly Loco Pilot, was shared with the legendary Joe Duddington, who was later destined to enter the history books by running "Mallard" down Stoke Bank at 125 m.p.h. As young Frank fired the ancient Stirling 0-6-0ST 611N round the Loco Yard, he was not to know that in the evening of his career he would follow Joe down the same tracks of the East Coast Main Line.

Before then, however, Frank had many years of hard graft ahead; with the Depression looming, any progress on the footplate could only be achieved by moving around, and after a few months at Doncaster Frank transferred back to Colwick. Work was still scarce, however, and although he had a rare day out on 1st August 1925 when he fired D3 4-4-0 1315N to Skegness, his more usual duties included cleaning ashpits, preparing or turning engines, or 'point holding'(?); sometimes six months would pass without a single footplate turn.

To make the most of what work there was, Frank moved to Derby for three spells in the Summers of 1928, 1931 and 1932, returning to Colwick between times; he then spent a year at Bradford, where his first job was the Laisterdyke Pilot. Back at Colwick for a spell, but still on 'odd jobs', Frank was eventually offered regular firing at Hornsey, and moved there in July 1935.

Here his horizons widened somewhat, with trips to such remote locations as Battersea, Herne Hill, Hither Green and Bricklayers Arms; there were other, more mysterious, entries in his diaries - what was the 'Caley Pilot' or the 'Crab', for instance?

Back in 1926 during one of the Colwick spells, Frank had worked a shift with one George Eaton, a driver from Langwith Junction. What, if anything, George told the young fireman about the place, we will never know, but years later an opportunity arose to exchange places with a Langwith fireman named Carter who wanted to move to Hornsey, and Frank relocated yet again.

Starting at Langwith on 21st February 1938, Frank was immediately thrown in at the deep end, with three straight lodging trips to Immingham firing to Charlie Maples, mostly on RODs but with one trip each on a 'Tiny' (Q4) and a 'Pom-Pom'(J11). The next week saw two more lodging turns, this time to Mexborough, followed by a day job to Darnall, all with Driver Bill Hulett, but Frank must have hit it off with Driver Maples, becoming his regular mate from the third week onwards.

Much of the work involved lodging, and sometimes Frank would work for several weeks solid on such jobs, with only the Sundays for a break. On Bank Holiday Monday, 1st August 1938, there was a change however; Frank fired J11 5306 from Chesterfield to Skegness and back on an excursion, booking 15 hours start-to-finish - a long day even if there *was* a break at the seaside!

At a lowly freight shed like Langwith there was of course no question of crews being allocated to regular engines, and Frank had a steady diet of O4's and J11's with an occasional 'foreigner' thrown in. There were a few North Eastern J27's allocated there at this time, and these would be used occasionally, along with the odd 'Sweedie' - Frank once took a J20 to Whitemoor, returning the next day on a J17, but this was exceptional.

It wasn't *all* main-line work even now; Frank renewed his acquaintance with George Eaton 'preparing engines' for a whole week, but there were no more days in the ashpit and Frank was steadily building up his footplate experience prior to becoming a Passed Fireman.

In September 1939 World War II began, but initially there was little change in the routine at Langwith. For a few months, Franks partnership with Charlie Maples was disrupted, and some weeks would see him working with a different mate and to a different destination every day, but things eventually settled down again, and the two men resumed working together until August 1940. Thereafter it appears that a new system was introduced, under which a Driver would change his Fireman every week or two.

A more far-reaching change was the abolition of Lodging Turns, due to the problems with food rationing as the effects of the War deepened. Franks last lodging trip was with an O4 to Gorton on October 2nd/3rd 1941, firing to Harry Allison.

Cancelling all the lodging arrangements must have been a nightmare for the scheduling clerks, and this is reflected in Franks diaries for the next few months. Apart from local jobs like trips to Pyewipe Junction (Lincoln), Frank again found himself going to a different place every day, though in part compensation the crewing arrangements changed again, giving him a regular mate for two or three months at a time.

In place of the long-established lodging system a new method of working was devised. This involved crews on Immingham or Whitemoor turns, for example, setting off in the usual way; in the light of their rate of progress, Control would instruct a signalman at one of the stations beyond Lincoln to stop their train at the same time as one going in the opposite direction. The crews would then simply exchange footplates and work back home again.

This resulted in a lot of new 'destinations' appearing in Franks diaries. Monks Abbey, the first suitable stop after Lincoln on the Market Rasen line, was a favourite, but if they were making good time or the opposing train was delayed, they could get to Market Rasen or even beyond before changing over. Similarly on Whitemoor trains, Woodhall Junction was a regular meeting place, but again Frank could get as far as Hall Hills (Boston) before being relieved.

Lodging turns to Gorton and Woodford were handled differently, in that the turn-round points would be Sheffield and Annesley, respectively, regardless of the rate of progress. Gorton trains were almost always double-headed, with a J11 piloting to Sheffield before coming off and banking the train as far as Dunford; this system continued, the train engine being re-manned at Sheffield and the pilot men working back from Dunford to Langwith either light engine or with return empties.

Frank fired for Bob Vamplew from October 1941 to the end of the year, and then had a spell with Jack Kidder until the next Easter; after that he stopped recording his mates name, which suggests that he was again working with a different driver each day. In February 1944 he became a Passed Fireman and two years later moved across the footplate permanently; this however put him at the bottom of the ladder again, and the immediate postwar period saw him spending much of the time on Shed Pilot or shunting duties.

In earlier years such a situation would have seen Frank looking for better opportunities elsewhere, but he was now married with three young sons, and his wife decided that Langwith Junction was good enough for *her* for a while! So Frank stayed put, biding his time.

January 1948 saw a small step in his career, with a full week on local passenger turns; Frank spent the time working between Chesterfield and Lincoln with a variety of C12's. Some idea of their condition may be gauged by the fact that although Langwith at that time had four of these venerable machines on its books, in the course of the week Frank had to use three loaned examples from Lincoln!

(26) *Frank Newham in uniform*

This week's work was presumably covering for illness or holidays, as it was another six months before Frank repeated the experience. Thing's hadn't improved - after two days with C12's, an N5 had to be borrowed on the Wednesday, a J11 on Thursday, and nothing less than Lincoln's B17 1667 "Bradford" at the end of the week!

At about this time much of the freight work at Langwith was being done by Gresley O2 2-8-0's which had temporarily usurped the usual O4's; ex-W.D. "Austerity" 2-8-0's were also beginning to appear, and Franks first experience of one of these was with 77453 on 28th May 1948. "W.D.'s" featured increasingly in the diaries over the next few years, even though Langwith didnt actually receive its own allocation until June 1954. By then the O2's had long gone, disappearing virtually overnight in June 1950 (Franks last trip was with 3946 to Lincoln on the 29th).

On Sunday 6th August 1950, Frank unknowingly reached another milestone in his career, when he took J11 64365 on an excursion to Skegness, returning with Lincoln B1 61329. Not a significant event at the time, maybe, but it was his first real taste of passenger driving, which would have a great effect on his later life.

One other turn in 1950 deserves recording; on 7th October Frank took K3 61944 on an unspecified job, handing over at "Midland Junction" (Manchester); he then lodged at Gorton (there must have been some facilities still remaining there for overnight stays), returning with the same train next day on a different K3.

Otherwise it was the "mixture as before"; a steady diet of coal trains halfway to Immingham or elsewhere, leavened by the occasional passenger trip. On the latter, Frank had his first experience of an A5 4-6-2T when he drove 69815 on 12th December 1950. In April 1951 Frank "looked over" the routes to Skegness and Cleethorpes prior to the summer season, and after a couple of weeks on standby for loco failures, his own turn came on 17th June when he took another Lincoln B1, 61279, to Cleethorpes via Gainsborough. Returning with sister engine 61329, he booked 16½ hours *that* day! Two weeks later he was again at Cleethorpes with a train from Kirkby headed by J11 64427.

(27) *Frank Newham and his fireman, the late Wray Morley, have changed into 'civvies' and gone for a stroll whilst laying over at an East Coast resort. c1958.*

1951 was the last year of operation for the line from Chesterfield to Langwith through Bolsover Tunnel, the line closing between Markham Junction and Langwith from 3rd December. Franks last freight trip on the line was with O4 63648 to Markham on 30th July, and his final passenger turn through the tunnel was on 15th September with 69812; he returned there a week after closure with an O4-powered ballast train, presumably to recover track before the tunnel collapsed altogether! Thereafter track-lifting was a leisurely affair; as late as March 1953 Frank had two trips to Scarcliffe with J11's. He also continued to reach Markham and Arkwright via Staveley.

The next job to disappear from the diaries was the Dunford pilot. Towards the end of 1953 electric locomotives were being prepared to take over the slog up to Woodhead from Sheffield, and bankers would no longer be needed; on 11th September J11 64321, a Langwith stalwart for over twenty years, brought down the curtain. Cancellation of this job seemed to virtually signal the end of mainline work for the faithful old Pom-Poms, their tasks increasingly taken over by the WD's; certainly the only subsequent entries for J11's concerned the Mondays-only Warsop Colliery turn and the occasional turn-out of the Breakdown Train. Since the latter would require the immediate availability of an engine in steam, this suggests that either a J11 was kept on standby, or more likely that they had replaced the N5's as Shed Pilots.

In place of the Dunford run, Frank began working with a 2-8-0 to Rotherwood, on the outskirts of Sheffield, where trains changed from steam to electric haulage. He also had hold of a Top Link engine for the first time on 28th October 1953, when on a special working to Marylebone he exchanged his K3 for V2 2-6-2 60915, probably at Leicester. Presumably he would have taken a conductor from there; at any rate he lodged at Cricklewood before coming back with the same two engines the next day.

1954 again saw regular excursions to the East Coast, often with B1 61405, a Lincoln engine from new; however Whit Monday saw a change with Frank taking LMS Crab 2-6-0 42786, probably with an excursion off the Midland.

June 1955 saw a brief railstrike by ASLEF, but unusually for a driver, Frank belonged to the NUR and stayed at work. This wouldn't have improved his popularity at Langwith, but he was confined for the duration of the strike to local workings to Thoresby or Welbeck Collieries; engine numbers were not recorded.

1956 saw a similar pattern to previous years; on August 25th Frank's mate on K3 61944 to Cleethorpes was Colin Palmer, who was destined for greater things, eventually retiring as Area Manager at Immingham years later. Frank meantime helped along his own career by learning the ropes as Relief Running Foreman, a job which figured at holiday periods over the next few years.

By 1960 Frank's sons were growing up, and he decided that the long-deferred move couldn't be put off any longer. On 11th January, at the age of 55, he moved back to Doncaster, scene of his first footplate experiences 36 years before. At a time of life when most Drivers would be looking forward to a few peaceful years of quiet shunting prior to retirement, Frank was busy learning the road, initially to Peterborough (Main Line and also via the Joint Line), York and, later, Leeds.

The contrast with his previous experience on plodding unfitted coal trains couldn't have been more complete - within a month of arriving at Doncaster he was taking A3 60066 to Newark! He also had to become familiar with other new types - 9F, A2 and B16 featured in his diary over the next month or two - though as another retired driver recently told me "they were all the same to us".

On June 5th Frank had what must have been a nostalgic trip when he brought K1 62046 to Langwith on an unspecified turn - it must have seemed like another world!

His subsequent career has no connection with the subject of this book, apart from the fact that his wife was unable to settle at Doncaster and they bought another house back at the Junction. Frank then had to commute daily, first on a moped and later in an Austin A35. However we have followed his progress for nearly forty years so will finish his story.

Frank had reached Doncaster just in time for the Indian Summer of steam on the East Coast Main Line, and for the next couple of years he happily blazed up to Kings Cross and back; he even started lodging again in a modest way, working up one day and returning the next on a V2 with the '112 Parcels'. Another regular duty involved running the "White Rose" both ways on alternate days, interspersed with a ride on the cushions to March to bring back a K3-powered goods.

These times were not to last, of course; the writing was on the wall for steam, and in September 1961 Frank reported to Ilford for initial training on diesels. After a few days he returned to Doncaster for static training on Brush D5815, followed by three shifts running to Hull and back; the next day he passed out on the new power, a far cry from the years spent accumulating steam experience!

The training courses continued apace - Deltics in November, EE Type 4's in December, and "Brush/ Sulzer's" the following January. In between, he had to revert to steam; on a typical week in May 1963, when coincidentally his mate was a certain Mick Newham, he took out an English Electric Type 3 two days, a Deltic the next, and finished the week on a 9F!

At about this time Frank's health, affected by long years on the footplate, began to suffer; he was off ill for the whole of June 1963, and again in October/November. The next year he was away again from mid-March to late May, and it was perhaps fortunate that most turns were now in the relative comfort of a diesel cab. One exception to this was on 1st October 1964 when he took 4472 (laconically marked "Peglers engine" in the diary) as far as Peterborough on the return leg of a Finningley excursion.

This turned out to be almost Frank's swansong on front-line steam; apart from a day as standing pilot with 60062 on 27th October, he only had one more Pacific turn, bringing A1 60155 back from York on 4th June 1965. Meantime he had another week with son Mick, a typical mix for the period with three days running to March with diesels, followed by three trips to Cleethorpes with a B1. On the latter there were no more 16-hour days - after reaching Cleethorpes, Frank and Mick went home on the cushions!

February 1965 saw Frank ill again, and he had more time off in September. By this time steam trips were becoming increasingly rare, though during week-commencing 22nd November he did a full week on a Doncaster-Leeds job which apart from the Monday drew a different LMS Jubilee 4-6-0 each day. On the Friday his diary is marked "Collision - Red Bank", but whether Frank was personally involved isn't mentioned.

Early 1966 saw all remaining steam at Doncaster withdrawn; Frank's last run on an Eastern Region type was on an early morning trip to York on 16th March, when he took B1 61319, returning with a diesel. After this there were three final trips on the 02.58 Doncaster - Leeds Parcels, which was rostered for a Holbeck Class 5 or, rarely, a Jubilee. For the record, the trips were :

17th March 44853 - Fireman Ralph Watkinson
19th March 45204 - " Barry Lane
12th April 45660 - " Paul Spence

As if to underline the changing times, Frank meantime took a Driving Course on D.M.U.'s, a poor investment as it turned out, since he did hardly any DMU driving afterwards.

During the summer of 1966 Frank drove the experimental diesel DP2 on quite a number of occasions as it worked up to Kings Cross, but there was little else of note for the next couple of years. The end, when it came, was quick; on 26th September 1968 Frank took a Deltic on the London run for the last time (D9007 Up, D9004 Down); a month later, on 23rd October, he was in charge of Brush Type 4 D1995 on the same journey. Next day he drove D5853 to Bardney, an echo of his Whitemoor changeover days with an O4 years before.

Then it was all over; Frank went off sick for the last time, and after three months away decided to call it a day. He took early retirement at the end of 1968, aged 63. Sadly, as for so many footplatemen, it wasn't a long retirement; on 2nd September 1973, he passed away.

16. LIFE ON THE FOOTPLATE (II)

(This song is from a collection on a tape given to the author by Mick Newham; the origin is uncertain but it is thought to have been written by a former footplateman at Kirkby-in-Ashfield (Midland) Loco. Most of the songs have local connections - "Last train on the Pinxton Line" and "Runaway Train to Toton" for example).

Fifteen years and five foot one,
Dressed in soot and coal,
Cleaning engines for a year,
Elbows deep in paraffin oil ....

Sixteen now and on the shift,
Carrying fire and sand,
Shifting clinker, raising steam,
Cleaning up and filling his can ....

Twenty-two and a Main Line man,
His trial has come and gone,
Fill the boiler, watch the gauge,
Open the door and shovel it on ....

Thirty-four and now he's driving
every other day,
but times get worse and the jobs go down,
And back he goes on firemans pay ....

Forty-eight and getting stout,
But the footplate's all his own,
His engine thunders through the night,
Spills his tea and rattles his bones ....

Five foot nine and sixty-one,
His glasses on his nose,
Can't see too well so off he comes,
To sweeping up and polishing floors ....

Sixty-five, his time is done,
He swap's his coal and oil
for a smart gold watch that bears his name ....

....then sixty-six and covered in soil ....

17. AFTER NATIONALISATION

Langwith being regarded as a railway backwater, there were few dramatic changes immediately after 1948; over a period of years the locomotives lost their 'L N E R' lettering in favour of, initially, 'BRITISH RAILWAYS' and subsequently the two versions of the B.R. crest. Smokebox doors began to be adorned by shed-code plates, Langwith being 40E until February 1958, when the code changed to 41J.

One matter which did, however, require urgent attention was the question of power for the passenger services. The old C12's, poor feeble things at best, were quite worn out and long overdue for replacement. A series of trials took place, featuring at different times a pair of 0-6-2T's from the London area, N2 69567 (at Lincoln 10/48 to 2/49), and N7 69694 (also at Lincoln 9/48); they even tried a "Tilbury Tank" - the local men thought *that* was even worse than the C12's, which was saying something! The two 0-6-2T's were strong enough, but their wheel arrangement didn't take kindly to the less-than-perfect local track, often affected by subsidence. This was emphasised when another N2, 69552, went down the bank near Mansfield as previously mentioned, so the C12's had to soldier on for a few more months.

Finally, early in 1949, help was at hand at last; displaced from London suburban duties by new L1 2-6-4T's, a batch of Robinson A5 4-6-2T's was rendered surplus at Neasden, and three of them, 9812 and 69815/18, were sent to Langwith. They were the ideal solution - big strong engines, fast and comfortable in either direction, and of course they had exactly the right pedigree for the local enginemen! Langwith never had much labour to spare in postwar years for cleaning purposes, but somehow they managed to keep the A5's clean and smart in their lined-out black livery. Apart from a brief period towards the end of 1954, when they were temporarily replaced by a brace of C13's, the A5's would serve until the end of local passenger services.

(28) *Class A5 69815 stands in Platform 1 with a Down train as 04 63707 heads north towards Sheffield*

On the shunting scene, the old N5's soldiered faithfully on, despite being almost sixty years old; later they were supplemented by a handful of J94 0-6-0ST's, and for a short time around 1959/60 the authorities even sent two or three of the diminutive ex-Great Eastern J67 0-6-0T's to help out.

As mentioned previously, a few B1's and K3's were drafted in, mainly for weekend passenger specials, though they were put on coal trains if power was short. For freight work, the exodus of the O2's has already been discussed; the replacement O4's plodded on, gradually being supplemented from 1954 onwards by WD 2-8-0's.

(29) *A long-time resident at the Junction, N5 0-6-2T 69284, pauses during shunting in front of the signal-box on 2nd September 1955. This old faithful arrived at Langwith in November 1939, and apart from a six-month stint at Boston in 1956 remained at the local shed until withdrawn for scrapping in February 1958.*

(30) *Having replaced the long-serving N5's on local shunting work, the J94's were in turn supplanted by Class 08 diesel-electrics. On 23rd September 1962 a grimy 68078 stands next to equally-filthy D3701 and D4089; the J94 was six months away from withdrawal. It was then sold to the National Coal Board, and may still exist in preservation.*

(31) *Three old warriors taking a weekend breather. WD 90301 is flanked by O4/8's 63765 and 63840, 23rd September 1962.*

On 1st February 1959, the neighbouring shed at Tuxford was closed, and its allocation, comprising twelve O4's and five J11's, was transferred to Langwith.

In 1965 a batch of 9F 2-10-0's, rendered homeless when Peterborough New England shed closed, arrived at the Junction; splendid engines for main-line work, they weren't at all suitable for venturing into colliery yards, and their reign at Langwith was brief, as was that of a batch of LMR 2-6-0's which also came from Peterborough. The last Pom-Pom had gone in 1962, so just the RODs and WD's were left to work out the final years of steam.

For the era of the diesel was coming; at Langwith it was at first no more than a handful of 350hp. shunters, then a few Brush Type 2's. They roughed it with the steam locos for a while, but a makeshift Diesel Depot was being fashioned out of the old Goods Shed at Shirebrook (West) on the nearby Midland line, and eventually the diesels moved down there; the writing was on the wall for Langwith.

The Diesel Depot opened in June 1965, and the steam shed was due for closure the following October, but as some of the replacement diesels were late on delivery, Langwith gained a short reprieve. However the 9F's and the 43xxx Moguls departed, mostly for scrap, and the last two O4's at Langwith, 63612 and 63843, were condemned in November (they would go to Cox & Danks at Wadsley Bridge the following January). On Christmas Day 1965, for the first time in perhaps seventy years, there was no locomotive in steam at Langwith for standby snowplough duties; instead a Brush Type 2 looked rather uncomfortable with an old Great Northern tender plough tethered either end!

(32) *A melancholy view of Langwith's last two J11's, 64379 and 64314, standing with tenders cleared waiting for their last journey to Gorton scrapyard. Taken on 23rd September 1962, when 64314 had completed sixty years faithful service. 64379 was two years younger, but had been at Langwith continuously since September 1939 - obviously not one of the black sheep!*

(33) *'Space-ship in trouble!' 9F 92195, after failing in traffic, has been towed back to Langwith with the r.h. con-rod lashed to the footplate. By this late stage in the steam era (1965), such a relatively minor problem would lead to immediate withdrawal, and 92195 was no exception; condemned in May 1965, and initially destined for scrap at Cleveland Dockyard, Middlesbrough, the sale fell through and she was still at Langwith on 30th August when this shot was taken. The following month she was re-sold to Drapers at Hull. Nearby can be seen one of the depot's collection of redundant coach bodies, a Gresley Corridor Brake-end.(E---49E).*

(34) *This view aptly illustrates the depths to which steam had descended in the final months of operation at Langwith. Taken in October 1965, it shows Driver Jeff Broughton and Fireman Stewart Whittaker oiling round a desperately neglected WD 2-8-0 90189 while the engine takes water - the Coaling Plant is just visible through the steam. Note the 41E (Barrow Hill, Staveley) shed plate on the smokebox door with the E roughly overpainted J for Langwith; the engine had just arrived and the local fitters obviously realised that it would not last long enough to merit a replacement plate. (90189 was withdrawn the following month).*

In the New Year the three remaining B1's were transferred to Stationary Boiler stock, leaving just a handful of WD's to see out the steam era. The survivors were by now in an appalling condition, with no spares available and little or no maintenance being done; as each locomotive failed it was simply laid aside to await a call to the breakers. By the beginning of February 1966 just three engines, 90153, 90572 and 90719, remained useable, and a week later, on February 6th, the last fires were dropped and Langwith closed to steam for ever.

By this time I was living in far-away Suffolk and it wasn't until February 28th that I was able to visit the Junction again. It was an unforgettable experience which I still remember quite clearly; as I walked unaccosted into the Shed Yard, I saw that although it was the middle of the day, every light was switched on, inside and out. In the offices, desks were strewn with papers as if the occupants had just slipped out for coffee; one workshop bench carried a pile of 41J shed-plates, the top one half-painted; on another was a pile of boiler washout books, neatly completed but with no entries after the previous June - whether the last book had been removed, or if they had simply stopped keeping records, I would never know.

In the Carriage Shed the last eight WD's stood cold and lifeless, awaiting their final journey to oblivion; outside, in a final irony, someone had neatly emptied and swept out the ashpits, perhaps for the first time. Somewhere a door banged, and I turned, half-expecting the old Shed Foreman to come running out with arms waving, but it was only the wind. The empty silence, after all the years of noise and movement which I'd come to know so well, was strangely unnerving, and I was glad to leave the place.

A few weeks later I was visiting the Junction again, and hearing an unfamiliar whistle, hurried down to the station. I was just in time to see an LMS Stanier 8F disappearing down the line; the signalman told me it had been onto the shed to turn. So the very last steam movement at Langwith Depot involved a type which I'd never seen there before.

On June 9th the last WD's were towed away for scrap and the Loco left deserted; some time later the Engine Shed was bought by Davis's for a Wagon Repair Shop and everything else swept away.

18. BRICKS, MORTAR AND STEEL

Now that every railway structure at the Junction has disappeared, apart from the railway houses and the shell of the Engine Shed, it may be worth recalling just what was there in steam days. The station itself was quite impressive, with four platforms each protected by half-canopies; the village was sufficiently proud of it to include a view of the station in a set of picture postcards sold locally in the 1930's. The main range of buildings was on Platform 1, the most southerly, which served the Down line to Chesterfield (on this east-west route, Down was arbitrarily chosen as westwards). The station building even boasted a Refreshment Room in the early days, but this had long been closed and was later used as a signing-on point for Guards.

The architectural style was unique to the LD&ECR, and comprised a series of modular bays, each 8'2" wide, separated by 22" brick pillars (31" at corners). Individual bays were built up in various ways according to the required interior layout - some were all windows, some had a door with windows either side, a few were all brick. One bay towards the centre of the building was left open for barrow/ pedestrian access to the station platform, though in later years this bay was filled in with wooden partitions to provide additional interior space. The number of bays at each station varied according to the anticipated traffic level - Langwith Junction had thirteen bays whilst Edwinstowe has twelve, Warsop nine and Clifton-on-Trent seven.

(35) *This recent view of Edwinstowe station building, which still survives albeit privately occupied, is included to illustrate details of the style of the one at Langwith Junction. and indeed almost all the LD&EC stations. The modular design is evident, although Edwinstowe has one fewer bay than Langwith; note the bricked-in section towards the centre, this was originally an open archway for pedestrian/luggage access. The squat chimneys are another characteristic feature of LD architecture, as are the deep-set low arches - note the freshness of the brickwork, protected until quite recently by the platform canopy.*

The buildings at the Junction, apart from the Goods Shed, had been razed by the time I began taking detailed notes, but the stations mentioned opposite still survive; from them a good idea can be obtained of the details of construction, with low inset decorative arches in each bay, along with squat bulbous chimneys and a lavish use of contrasting blue brick.

The central island platform at the Junction had only a simple wooden partition with back-to-back seats under its canopies, but there was a small range of brick buildings, seventy-two feet long, on Platform 4. The depth of these was restricted to eight feet as the end track of the Up Yard lay immediately behind; apart from a little-used waiting-room there was an office for the Assistant Station Master, a position which survived until after WWII. The last incumbent rejoiced in the name of Onions, but this was obviously no handicap to his career, as I understand that he finished up as Station Master at Liverpool Street! The buildings on Platform 4 were out of use by 1957 and had been demolished by 1960.

Linking the platforms at the east end was an ornate lattice-steel footbridge, decorated with wrought-iron work and cast-iron flower heads; it was extended at either end right across the railway property (nine tracks plus the Goods Yard at this point) to provide access to a public footpath over the fields to Nether Langwith. This footbridge became life-expired in 1960 and was replaced by a more utilitarian structure erected a few feet to the east; the big 75-ton crane from Colwick had to be borrowed from another job at Warsop to lift the spans into position, the local crane's jib being too short for the task. The new bridge had solid sheet-steel sides rather higher than the original, which made life more difficult for train-watchers!

(36) *N5 0-6-2T 69263 shunts excursion stock in Newtons Sidings, 20th August 1960.*
Both the engine and the footbridge, part of which is visible in the left foreground, have only months left before scrapping. Note the concrete-post signals, also the patch where the station building has recently been demolished.

The platforms themselves were edged with stone flags $4^1/2$" thick, each flag being 36" wide and varying between 56" and 63" long. The facings were built in blue brick and were 35" high; curiously the pattern of the overhang had three steps on Platforms 1,2 and 4 but only two on Platform 3. The platform surface was ashphalt adjacent to the buildings but ash elsewhere. There were the usual flower beds edged in whitened stones - as with most passenger stations the staff took a great pride in their "gardens".

Just off the east end of the centre platforms stood the Junction signalbox, a handsome brick structure containing a Saxby & Farmer tappet-locking lever frame on its south wall. The frame had originally comprised 81 levers, but around the time of the First War, a small box at Langwith Junction (East), which controlled the western end of Warsop Loops, had been demolished, and four extra levers added to the main box. These were placed at the end of the frame next to the existing No. 1 lever, so to avoid renumbering all the levers, the new additions were marked A-D! The box had to be slightly extended to accommodate the lengthened frame, but this was done so neatly that there was little evidence of the alteration, the only clue being the extra windows, which were 6" narrower than the originals.

Langwith Junction Box had a Class 1 rating, the next to the highest in importance, and for many years was open continuously seven days a week; towards the end it was closed between 6 a.m. and midnight on Sundays.

(37) *A view of the signal-box taken c1974 after the Chesterfield lines had been lifted. The pair of slightly narrower windows at the nearer end are the only visible evidence of the lengthening of the box carried out to accommodate extra levers when the old East Box was closed. The ugly replacement footbridge in the background has none of the elegance of its predecessor.*

One of the functions of the old East Box mentioned above had been to work the two sets of points giving entry to the Loops, and as these were several hundred yards from the main box it was impractical to provide manual linkage; instead an early design of power points was installed. The primitive motors used were somewhat unreliable, particularly under conditions of snow or hard frost, when they would grind ineffectually away for a while before expiring in a cloud of smoke! This usually seemed to happen in the middle of the night, and we would then be awakened by a tapping on the drainpipe and a gruff voice calling *"Electric points, Len"*. That was all that was necessary for Dad to leave his warm bed and spend the rest of the night barring the points over by hand; it made a change, however, from the even more numerous occasions when the voice would shout *"Foggin', Len"*, which entailed him putting detonators on the rail at the foot of some signal for several hours. Needless to say, he was expected to be available for work at the usual time the next morning, though he was allowed home, briefly, for breakfast!

Returning to the station environs, there were the other facilities usually associated with a medium-sized country depot - a brickbuilt Goods Shed with adjacent hand-powered crane, a cattle dock with an open loading bay alongside, an end-loader, and a separate track with access alongside for road vehicles. This latter had for years seen no goods traffic, and was invariably used to park Goods Brake Vans, being convenient for engines going off shed to collect a Van as they departed. The open loading bank was edged with special blue bricks, which featured a non-slip chequered top surface and a rounded edge; occasional bricks were also embossed *"Joseph Hamblet, West Bromwich, 1895"* (some were *1896*). By sheerest chance many years later, after the site had been cleared, I came across one of these bricks half-buried under a pile of rubble, *and* it was one of the embossed ones! It is now a valued part of my collection.

(38) *K3 2-6-0 61950 heads purposefully for Lincoln and Cleethorpes with an excursion, 9th June 1957. The train is crossing the bridge over the Midland line just east of the Junction; on the right are the 'motor points' giving access to Warsop Loops. The former Up 'New Found Out' link to the Midland route had previously diverged from the Lincoln line under the fourth coach before swinging to the left behind the nearer bracket signal.*

The Signal & Telegraph Department had a wooden store alongside the cripple road off the Up Yard headshunt, and they also used for storage an old covered van of Great Eastern origin; it still bore plates marked "Stratford 1912". After lying about the Junction for years, it finished up parked at the end of the Brake Van siding, where I believe it was finally broken up, too far gone to move.

The original LD&EC signals were I believe of somersault pattern like the GNR, but these had been replaced by ordinary lower-quadrant examples between the Wars. Later still these in turn gave way to upper-quadrant types, some on concrete posts, but the replacement process was a slow one. The tall Down Homes off the Midland connection lasted until around 1950 when those tracks were removed, and the Up Starter on Platform 2, a right-hand bracket type, was the last survivor when replaced in October 1952.

19. THE VILLAGE

(39) *Three more examples from the postcard collection sold locally in the 1930's.*
(above) Chatsworth Avenue (on right) and The Crescent (on left). The authors childhood home is the furthest house on the left.

(40) *(above) Recreation Road, with council houses on right and the first private (£400!) houses on the left - more of these would eventually fill the gap behind the field fence. The splendid gas lamps were replaced in the '50's when they could be purchased for £1 each - what price now?*

(41) *(above) the main road through the village, with Burlington Avenue on left and Harkers Cottages opposite. The Post Office stands on the corner of East View, which leads to the station; apart from the wooden canopy over the shop windows, the present scene sixty-five years later is identical (apart from the traffic!).*

Although the railway has almost entirely disappeared, the village of Langwith Junction remains remarkably intact; indeed apart from a few "pre-fabs", *every* house built since the 1890's still stands.

Originally a green field site with only a minor road from Shirebrook to Langwith passing through, the Junction began with houses built by the LD&EC for its employees. These comprised a substantial detached house for the Station Master, strategically placed overlooking the Goods Yard, and a short terraced row plus a pair of semi's, which for some reason were built several hundred yards away alongside the tracks towards Chesterfield - perhaps it was thought that this site would be quieter than one near the station! All these were built in the same 'house style' as the station building already described, but another short row near the Goods Yard was for some reason built to a much plainer design. This latter row had (and still has) the unusual 'distinction' of having no direct road access, just an unmade track along the rear of the houses built subsequently along the main road.

The main-road houses comprised several long terraces and a selection of what in those days were called 'villa's', mostly in pairs, which were privately built soon after the railway was opened. One long row rejoiced in the name of 'Harkers Cottages', presumably after the builder (or landlord). There was another short row, East View, at right angles to the main road and alongside the pedestrian approach to the station, and in the right-angle thus formed a Village Store and Post Office was built. Later a couple of Harkers Cottages had their 'front rooms' converted into tiny shops, but it wasn't until about 1930 that the village really began to expand, with the addition of several streets of quite substantial council houses, a row of detached houses (priced from £400!) and a Co-op. For a pub the Junction had to wait until postwar years, though there had been a licenced club since early days.

Right up against the entrance to the Loco Depot and well away from the rest of the village was a short cul-de-sac called Eland Road; this inevitably grimy area was rather looked down on by those of us who lived at the "top end"!

Just after WWII a rash of prefabs appeared in the fields to the west of the village, followed by a further large estate of council houses, making the Junction quite a substantial village.

20. THE MISSION

One building which sadly *hasn't* survived is the Mission, as the local 'tin tabernacle' was always called.

When the LD&EC was first opened, the brick Goods Shed was used as a place of worship by railwaymen, but as business developed this soon became impracticable, and the patriarchal LD&EC General Manager, Harry Willmott, was approached by the churchgoers leaders to see what could be done. Mr. Willmott arranged for the gift of a plot of land alongside the railway, adjacent to the station footbridge, and also donated the necessary funds for the erection of a simple but adequate corrugated-iron building, to be used for worship on Sundays and as a Social Club and Institute during the week. The building was first occupied in 1902, when the resourceful Mr. Willmott instructed each station on the line to donate one of its platform benches to the Mission, in lieu of pews! In those days the station name was painted on the back rail rather than being embossed on a cast plate, and for years afterwards the original lettering could be seen under the successive coats of paint! When additional seats were required in the '30's, some closely-slatted long wooden benches arrived; it was rumoured that these came from scrapped Underground trains! Whatever their origins, they were certainly more comfortable than the old platform seats!

(42) *The Mission celebrates the Royal Coronation, 1937.*
Left to right : Albert Preston, Mesdames Mugglestone (Harold Watts' sister)/-/Betts/Wytcherley/H. Watts/Preston/Andrews/ T.Wilkinson/J. Watts/ S. Wilkinson/Saxby/-/Pogson, Harold Watts.

(43) *The Mission interior, on the occasion of it's Golden Jubilee in 1951.*

The Mission was always a spartan place, with little in the way of decoration. One item I do remember, however, was a large framed painting which hung on the left-hand wall near the front; it was entitled "The Broad and Narrow Way" and depicted the view from a high place, with two alternative paths leading away. As can be imagined, the Narrow Way led through pleasant gardens past church and Sunday School on the way to a Golden City on the skyline, whilst the Broad Way was smooth and well-paved, passing various temptations including a tavern, theatre, Ballroom and Gambling House, even a Lottery (shades of 1995); beyond, there were muggings, fights, and a full-scale war before the Broad Way ended in the pits of everlasting damnation. What has remained in my mind, however, is a small sketch of a green 4-4-0 and a string of coaches in the far landscape entitled "The Sunday Train" - needless to say, it runs by the Broad Way!

The Mission, like most places of worship, thrived until the 1950's, with more than a hundred scholars in the Sunday School. (Secular activities had been transferred before the War to a new Social Club, built near the Loco). Then a long slow decline set in, culminating in the sad decision, taken at the Service held on October 1st 1989, that the dwindling congregation could no longer support the Mission. The few survivors joined a Methodist Church at nearby Shirebrook, and the weary old building was left deserted, finally being pulled down as unsafe in 1993. The platform seats had previously been sold off cheaply, though they are now in great demand as collectors items!

21. PERSONALITIES

Jack Mallett, previously mentioned as a source of information on Links and Lodging, also added some names to my own recollections. Prominent amongst the drivers, who in those days were always regarded as the elite among railwaymen, was pensioner Ernie Hulett, whose claim to fame was that he was the last surviving LD&ECR driver still living at the Junction; his home was in The Crescent, a few doors from that of the writer - he was a living link with nineteenth-century railways!

(44) *A c.1921 group photo of the Mutual Improvement Class and its just-acquired mobile classroom, GCR Full Brake 1654A. This vehicle, which contained seats for 27 pupils as well as a lecture platform and display of instructional exhibits - injectors, lubricators etc - remained in use until 1953.*

From left to right (front): -/-/Billy Deakin/George Yates/Edward Kemp/Tommy Booth (with watch)/Harry Ayres (head only)/-/Len Jakeman (head only)/Bill Ayres (with stick)/-/-/Ray Taylor/Charlie Maples (head only)/-/-/-/Jack Gilbert/Lefevre/George Grainger/Ernest Bradshaw (chargehand-fitter)/Smith (instructor)/Bill Newton (Foreman)?. Back - second from right : Oliver Harrison. Note the 'Tiny' (Q4 0-8-0) on right.

(45) *The original print of this shot shows the O4 tender lettered LNER with the number below, which dates the view as between 1923 and 1929.*
Left to right (bottom row): Percy Bearpark/George Lymm/Bill Palmer/Les Lee (still alive at time of writing)/George Wakefield/ Fred Dutton (lathe-man)/Jack Booth (Union)/ -/Nelson Bollington/Jack Martin/Bill Mayes/Obadiah Taylor/Wilf Roberts/Billy Green No.2/-/Harold Houghton/-.
(top row): Bill North/-/Frank Cartwright/-/-/Tommy Derney/B.Palmer(?)/Matt Lowe.

Other names from pre-1939 days include Albert Preston, who was Sunday School Superintendent at the Mission, and George Grainger, who assisted Albert there and later took over when Albert retired. It was George who gave me my first footplate ride, on C12 7357, through Bolsover Tunnel around 1947.

Two drivers who lived either side of us in The Crescent were Ted Cooke and Harry Qualters; Bob Cartwright and Stanley Heafield were other drivers living nearby, along with Frank Goodson, a genial giant who shared Dad's hobby of breeding canaries; I spent many hours sitting on the steps of Frank's garden aviaries, listening to his yarns about footplate experiences - if I'd had the sense to take some notes I could have written another book!

Then there were the Childerley brothers, Claud, Arthur and Reg, and Bill Hulett, Ernie's son, who later became a Chief Inspector at Doncaster. Family loyalty to the railway was very strong; apart from the Childerley's, the Frisby family, father James and sons Alan, Sidney, Ivan ("Fuzz") and Arthur were all drivers at Langwith.

Geoff Lacey was a Passenger Guard, always dapper with a flower in his buttonhole, and Driver Harry Allison (another Mission teacher) and Goods Guard Fred Garfitt also lived nearby. Tommy Bates, the Permanent Way Inspector, was Dad's boss, and Harry Morphus was a fellow lengthman. Mr. Gasson was the Station Master - I recall his waxed moustaches, just like Hercule Poirot! (Mrs Gasson, who we knew affectionately as "Fanny", was Senior Mistress at the local school); when Mr. Gasson retired his replacement was John Parker - a rather stern, unbending individual, he was always " Mr. Parker" to everyone.

22. THE WAGON WORKS

The only opportunity for employment at the Junction apart from railway service was the wagon building and repair shops of W.H.Davis & Sons; their Works lay behind the Loco and alongside the Leen Valley line. Davis's was an old-established concern which in pre-war days had built up a wide-spread network of repair depots for wooden wagons. Years ago I visited their premises hoping to discover more about the history of the firm, but found them totally uninterested in the past, being perhaps understandably more concerned with an uncertain future. I wasn't even able to copy their meagre stock of photos, though I did sketch some details, mostly of R.C.H. 7-plank opens, with examples built for Sheepbridge, Bolsover, Shireoaks, Welbeck, Alfred Blackmore of London, J & G Wells (Eckington), even some lettered LNER. They had an old view of the erecting shop with a large number of wagons for Hall & Co in various stages of construction. The only steel wagons which had been recorded were a pair of "Iron Mink"-style vans for Blue Circle Cement and Ferrocrete respectively, but whether these were overhauls or new-build I was unable to discover.

Certainly the firms fortunes seemed to decline along with those of the wooden wagon, and their position was not of course helped by the more recent decision of British Rail to restrict wagon work to their own workshops. This left Davis's and other private concerns solely reliant on work from the private sector, and wagons were built for London Transport and various quarry owners in later years. Davis's were also responsible for the well-known Milk Marketing Board "white-elephants", a series of refurbished milk tank wagons comprising new stainless steel bodies on second-hand ex-oiltanker chassis which were built, stored for several years, then scrapped, probably without ever carrying milk!

(46) *A general view of the Wagon Works which appeared in Davis's advertising manual - undated but probably late pre-war; the LMS wagon with small lettering visible right of centre would be post-1937. The Works Yard is suspiciously tidy, probably retouched! Note the grounded coach bodies, used as workshops, behind wagons on right - these lasted until*

(47, 48) *Two interior views of the Wagon Works from the same advertising manual. One shows the main Repair Shop (note that several wagons are lettered WHD&S on the ends), the other the Wagon Building Shop, with a large batch of 12-ton.*

the Loco Depot closed in 1966. The new turntable would later occupy the bottom right-hand corner of this scene. The curved-roof building on the right was reputed locally to have previously been a WWI aircraft hangar! Houses in Eland Road can just be seen extreme right, and St.Joseph's R.C.Church is also visible in centre background.

7-plank R.C.H.-pattern coal wagons under construction for Hall & Co. Again the foreground looks unusually tidy!

In recent years Davis's has had to reorganise several times in order to stay afloat, and although a certain amount of rail-connected work is still carried out, their main occupation these days is the manufacture of shipping containers (which sadly are despatched by road). When the Loco Depot closed,they bought the site and the Down Yard; the latter is still used for wagon storage, and the Engine Shed, much rebuilt, serves as a workshop, allowing much of the original Works to be sold off for non-rail purposes.

At its peak the Wagon Works covered several acres, but I never saw a Works Shunter there in steam days. Wagons were moved around by capstan, although there was also an isolated track at right-angles to the fan of sidings, which was home to a curious vertical-boilered contraption, probably built at the Works, which fussed up and down towing a one-wagon transporter truck. In recent years the Down Yard has held various relics including a Motor-Rail petrol shunter and a small ex-Army diesel, but they seemed to be rarely used.

A recent visit led to the discovery of a hard-backed promotional book which Davis's had at one time issued for publicity purposes; it is undated, but from the photographs appears to have been printed in the late 1930's. It lists thirty Depots for wagon repairs, ranging from Birmingham to Bristol and from London to Peterborough. Many of the photos seen on the earlier visit are included, with additional ones showing wagons built for Rose, Smith (London), Tottenham District Light Heat & Power, Chandler & Son, and Renwick, Wilton & Dobson Ltd.

23. LIFE ON THE TRACK (I)

Dad started on the railway in 1919 after service in the first World War, the only other job available locally being "down the pit". He spent the next forty years working at the Junction as a Permanent-Way Man; quiet and conscientious, he was offered promotion several times but always refused it, content apparently to work on the track. For many years he was a Lengthman, allotted a certain stretch which he patrolled daily with a hammer over his shoulder to knock in the loose keys used on bullhead track; he also took an oilcan if there were points to lubricate. It was a responsible job with precious little reward; carelessness or inattention on his part could soon have led to a derailment.

To supplement his meagre wages, Dad often spent Sundays working in the noisome depths of Bolsover Tunnel, where a constant battle was waged in an ultimately doomed attempt to keep the line open. His reward for this just covered the purchase of a 15/- (75p) Savings Certificate for the writer!

In later years he became responsible for driving the Platelayers Trolley, a rather primitive contraption which lived in a wooden shed alongside the P.W.hut next to the Up Yard Headshunt. The trolley was a standard Wickham-built model, with front and rear screens and a curved roof but completely open sides. Back-to-back wooden seats were fitted down the centre of the trolley, providing spartan accommodation for five men plus the driver; tools, materials and the rest of the crew would be carried on one, or occasionally two, small flat trailers which were hooked on behind. The trolley was powered by a big Vee-twin JAP engine, a development of the one used in speedway bikes I was told, and the drive included a large vertical disc which presumably formed part of a semi-automatic transmission.

When the ensemble reached the location of the days work, the trailers would be unloaded and lifted bodily onto the lineside; the trolley however, was too heavy for such treatment, and use would be made of little run-off tracks at right-angles to the line (a very common feature but one rarely recorded or modelled). In order to reach these tracks, the trolley carried a "portable turn-table", a very ingenious device comprising a pair of light metal bars, each flattened at one end and turned up at the other. These were laid on top of the running rails in front of the trolley and connected by a cross-piece, the latter being drilled in the centre so that it could be pivoted on a steel upright driven into a sleeper opposite the run-off. The trolley was then driven or pushed onto the bars, swung round through ninety degrees, and rolled off clear of the line.

Incidentally, the upright steel pivots in the sleeper centres were always covered by a conical wooden block when not in use to avoid an unwary boot tripping over them - an early example of Health & Safety at Work!

Apart from transporting the P.W. gang around, Dad's other task with the trolley was to deliver drinking water in milk churns to the level-crossing keeper at Norwood, about a mile down the Beighton Branch. This was a daily job since his only source of water, a well in the back garden, had been contaminated. Sometimes I would be allowed to ride with Dad on the short journey, but I was always instructed not to join him until he had reached the deep cutting beyond Newton's Yard, where we would be out of sight of any roving Inspector! The noisy engine, open sides and very low riding position gave a great impression of speed, out of all proportion to the very modest performance achieved at best - rather like driving a Mini!

Dad stayed on at work for several years after reaching the age of sixty-five, but eventually retired a few weeks short of completing forty years service; a grateful British Railways thereupon told him he didn't *quite* qualify for a gold watch! They did however grant him a pension, amounting to 30/- (£1.50) a month!

(49) *My father Len Little as most people remember him, with his inseperable companion "Rikki". c.1970.*

24. LIFE ON THE TRACK (II)

(Another song from the "Kirkby Collection")

I'd like to be a Lengthman, along the iron way,
Living among the fishplates, the sleepers and the keys,
With me forty pounds of hammer, if me accent and me grammar
were a little bit on the rough, they wouldn't mind,
so long as I keep attending to the line.

I'd like to be a Lengthman, a section of me own,
I'd keep it in good order, the best that I could do,
If they want the station shifting, or the track should need a-lifting
on a Sunday afternoon I wouldn't mind,
so long as I keep attending to the line.

I'd like to be a Lengthman when winter's on the line,
I'd volunteer for duty, whatever the time o' day,
and all that I'd desire, is a kettle on the fire,
for a couple of cans o' tea I wouldn't mind,
so long as I keep attending to the line.

I'd like to be a Lengthman, working with a gang,
in a wagon-load of ballast I'd be pleased to do me share,
If me wellingtons I'm wearing, and the culvert needs repairing,
and they keep us after time I wouldn't mind,
so long as I keep attending to the line.

I'd like to be a Lengthman, all on the iron way,
for Company and Country, I'd labour every day,
and I could always take a notion, for to try me with promotion,
for to wear the foreman's suit I wouldn't mind,
so long as I keep attending to the line.....

25. THE FINAL YEARS

My strongest memory of those early post-war years was the feeling of *permanence* of everything connected with the railway scene - the old N5's, or machines very like them, had been shunting the yard for over fifty years, and most of the local engines had been built in the early years of this century; it never even crossed our minds that things would ever alter. But alter they did, and very soon; subsequent changes in the locomotive scene have already been described, and the railway network was soon to change too.

The first contraction was the removal of the connection down to the Midland line through Shirebrook (West), already mentioned under signalling. This link, a west-to-south flying junction with the Worksop-Mansfield line, was always known locally as the "New Found Out" for some long-forgotten reason, and had been almost disused for several years prior to its removal around 1950; it probably owed its survival latterly to potential value as a diversion in the event of bomb damage to other routes during the War.

ROUTE CHANGES AT LANGWITH JUNCTION 1968-74
N
CLOSED
9·1·67
LDECR
MR
NEW LINK
9·1·67 TO 11·11·74
CLOSED
8·9·69
CLOSED
11·11·74
NEW LINK OPENED
11·11·74
CLOSED
1951
LDECR
CLOSED C1946-50
REINSTATED
11·11·74
BRIDGE
REMOVED 11/74
CLOSED
C1946-50
SHIREBROOK
WEST
CLOSED
27·5·68
GNR
MR

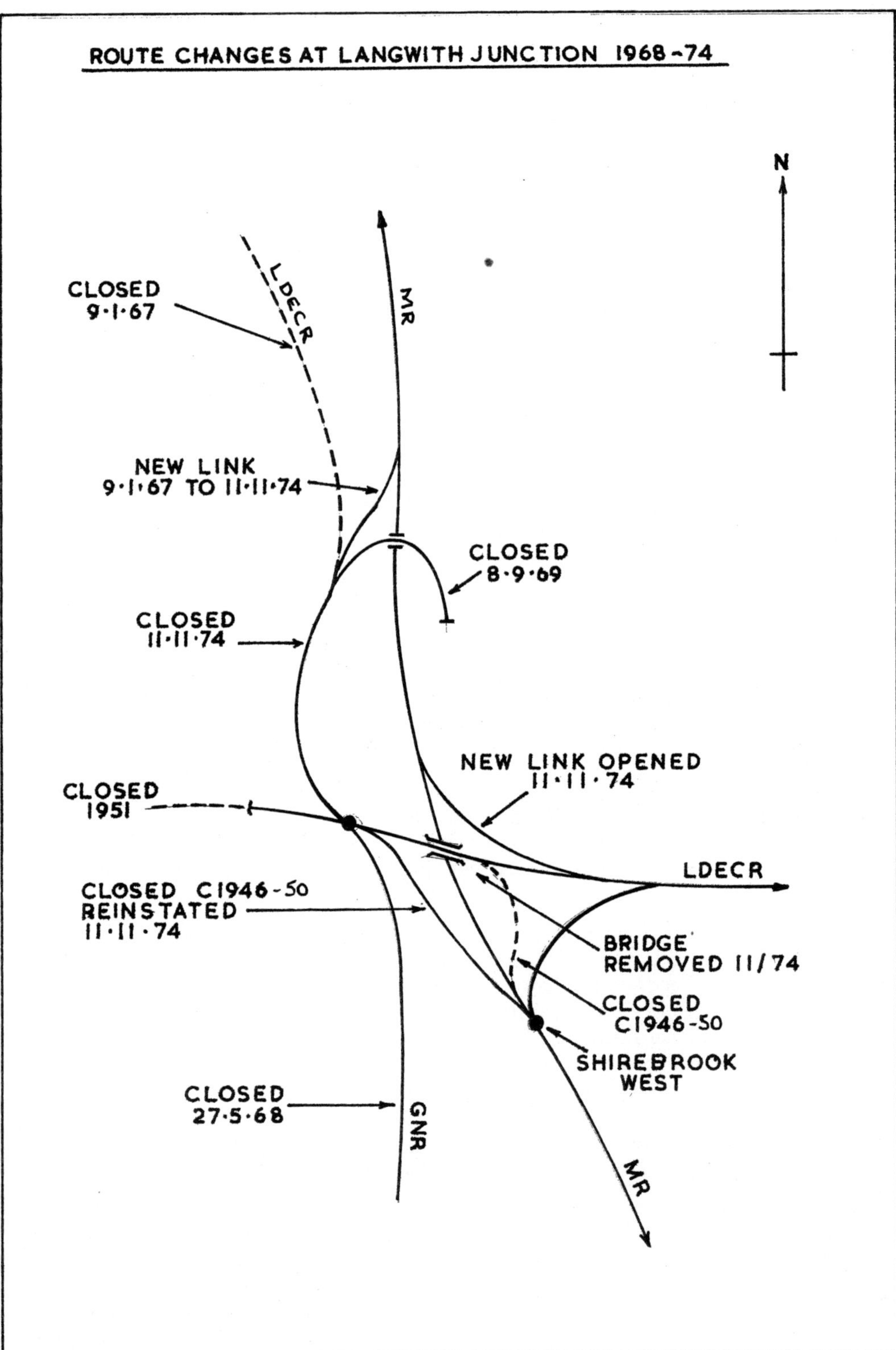

DIA.4

Much more significant was the final defeat of efforts to keep open the subsidence-ridden bore of Bolsover Tunnel, to the west of the Junction. In the 1930's the LNER had been given the option of buying out the coal reserves lying under the tunnel, but had declined on the grounds of cost, a decision subsequently regretted, as the expense incurred in maintaining the tunnel in later years was far greater than the value of the coal. Dad and his mates spent most weekends fighting a losing battle trying to keep the bore open. Being over $1^1/2$ miles long and on a continuous gradient of 1-in-120, Up trains had to work hard throughout the tunnel so it was always full of smoke; working conditions can only be imagined! At one stage the desperate measure was taken to reinforce the bore with lengths of rail, each curved into an inverted horse-shoe with the open ends sunk into two feet of concrete laid in the cess; the rails were placed at 12" intervals over a considerable distance, with old sleepers packed in horizontally behind, but even this extreme measure only delayed the end for a while. The bore continued to shrink until single-line working had to be instituted with one track slewed towards the centre; finally even this became unsafe, and the line closed completely between Markham and Langwith Junctions in December 1951. Over the following period the tracks were slowly lifted, leaving just a few hundred yards west of the Junction platforms, which were retained for occasional use as coach or wagon storage.

The truncated Lincoln service only lasted another four years, although some seasonal and excursion traffic ran for a while longer. In 1963 the Goods Shed was closed; large doors were put in one end so that it could be used to garage a railway lorry for a time, it then stood empty and derelict long enough to have its measurements recorded.

The next change affected the Beighton line. Near Norwood Crossing, mentioned earlier as being waterless, the line ran parallel to and at the same level as the Midland line from Worksop to Mansfield and Nottingham, with only a fields-width between; by building a short connection across the field, it would be possible for the LD&EC line north of this point to be abandoned. This was duly done, and from 9th January 1967 the new connection came into use; there would be no more Diversions.

The next closure concerned the Great Northern line southwards from the Junction, officially the Leen Valley Extension; this was reballasted in early 1968, only to be closed completely on May 27th of the same year! Then, on September 8th 1969, the short branch to Langwith Colliery was lifted.

By this time, of course, the Steam Shed at Langwith had been closed for some time, and the Junction was only a shadow of its former self. But worse was yet to come; the through-girder bridge by which the LD&EC crossed the Midland line just east of the Junction was declared life-expired (I'd crossed it on foot some years previously and noted the ragged state of some of the steelwork). The planners decided that since all traffic crossing the bridge was destined to join the Midland line north of the Junction, it would be more cost-effective to put in a east-to-north connection adjacent to the bridge which would then not need replacing. This involved a deep cutting on a severe gradient, but was eventually completed and from 11th November 1974 was opened to traffic, rendering all the remaining track at the Junction redundant overnight.

Well, not *quite* all - the Wagon Works was still in business, and although most of its work was by now concentrated on international shipping containers and other non-rail orders, there was apparently still sufficient wagon repair and construction work to justify maintaining a connection. The solution was to reinstate part of the long-removed "New Found Out", the Down Side link to the Midland line, and so the young trees were removed from the empty trackbed and rails relaid after a twenty-five-year gap. (Coincidentally, another missing connection to Shirebrook (West), that to Warsop Colliery, was also replaced about the same time, even though it ran through the new Diesel Depot!).

(50) *Conversation Piece : J11 0-6-0 64379 shunts the goods yard, 20th August 1960. By this date the Goods Shed was little used and the wagons are probably being stored awaiting entry to the Wagon Works. Note the old Pullman Car and the ex-Great Eastern Stores Van parked in the Brake Van siding.*

(51) *The last passenger train from Shirebrook North to Lincoln, 17th September 1955. A5 4-6-2T 69828 waits to depart with a specially-strengthened train comprising the usual Thompson 3-coach set plus two GCR veterans, added for the anticipated last-day crowds - there were none!*

(52) *Class J11 64389 takes the Beighton Branch with a Sheffield-bound excursion, 28th April 1956.*

(53) *Class O4/8 63801 heads through Platform 3 towards Sheffield with a Down empties, 26th August 1961. The new footbridge stands in the background; the platform gardens are sadly unkempt, six years after passenger services ended.*

(54) *A bird's-eye view of the west end of the Shed Yard, August 1954. Two of Langwith's three A5's are standing in front of the shed, and a Q1 0-8-0T is just inside. A visitor from March, K1 2-6-0 No. 62055, is in the foreground, and it is interesting to note that apart from this engine and a couple of WD's, all the locomotives present are of Great Central origin, thirty years after Grouping! The O4 on the extreme right is standing on the site of the original turntable.*

As already mentioned, Davis's took over the Down Yard and the shell of the Engine Shed but apart from these and a short section of the Beighton line retained as a headshunt and run-round for the Wagon Works sidings, all else was soon swept away - the station buildings, the signalbox (though not until it had been thoroughly vandalised - a heart-breaking sight for those of us who remembered its immaculate brasswork and polished linoleum), cattle dock and the rest. The platform edging slabs were removed, though the platforms themselves, swept bare, remained for a few more years, as did the "Cenotaph" coaling plant, which stood stripped of all metalwork for a while until finally being demolished to make room for a new sky-crane to handle the shipping containers.

Eventually someone realised that the modern footbridge was now only spanning a headshunt, so the line was shortened to be clear of the platforms, and the bridge carefully dismantled for use elsewhere. The truncated tracks was still long enough for Davis's modest needs, and even saw a Brush Class 47-hauled excursion a few years ago!

About 1985 the station site, along with the remains of Newtons Quarry, was sold for a new luxury housing development, but the ensuing depression saw this plan curtailed and only a few houses have actually been built.

The rest of the Junction remains much as it did twenty years ago, the few remaining tracks overgrown, rusty and rarely used. The whole scene is in total contrast to those hectic days which a few of us remember from the '40's and '50's, a time when Langwith Junction was busy enough to merit a comment in a railtour brochure likening it to Clapham Junction!

(55) *The last passenger train to visit the Junction was this railtour, the "Dukeries Anniversary", which is seen struggling up the reinstated "New Found Out" top-and-tailed by Class 47's 47489 and 47844 on 24th November 1990. The severity of the gradient can be judged by comparison with the Midland line to the left, which the train has just left. The pristine-looking tank wagons are in fact rain-soaked relics awaiting scrapping by Davis's.*

(56) *Langwith Junction today: a 1994 view of the loco yard, now used by Davis's for wagon repair and construction. The right-hand building is the former engine shed, re-roofed and fitted with new ends. The two bogie covered-hopper wagons in front of the shed are newly-built JGA's 17455/56, destined for Cleveland Potash. To the left the Down Yard is filled with private-owner wagons awaiting repair or scrap; the reinstated Down connection to the outside world is just visible extreme left.*

26. POSTSCRIPT

As small children, we were always allowed to stay up late specially on 31st December, to "see in the New Year". Just before midnight we would go out into the back garden - it always seemed to be a clear, starlit night - where we would stand and shiver in the darkness. Then, on the stroke of midnight, a whistle would sound from the direction of the Loco half-a-mile away, and at once there would come a great chorus of sound as every engine on shed added its contribution. The clear soprano of the LNER designs soared above the mellow contralto of the Great Central engines, with the hoarse hooters of the WD's providing the bass. The only thing missing was the chime of an A4!

After a minute or two, the chorus would slowly die away, and off we would go to bed, happy that everything was fine for another year. Some of the engines we had heard would have been observing the same ritual since Queen Victoria's time, and we had no reason to think that it would ever be any different. Little did we realise.....

(New Years Eve at Langwith Junction is nowadays as dark and silent as any other night, but they do say that if you listen very carefully.....)

(57) *A visiting K2 2-6-0 from Immingham stands on the ashpits, 22nd May 1950. The Wagon Works is visible on the right.*

APPENDIX 1 : Langwith Shed Allocations

1st January 1923*
(*All numbers transposed to LNER 1923 series for purposes of comparison with later years, except Class J58 No. 66B, which was scrapped in 1923 before receiving its new number).

C14 : 6120/21/23
G3 : 6402-07
J11 : 5204/17/18/22/23/29/39/40/46/56/57, 5312/17/18, 5948, 6080
J58 : 66B*
J60 : 6408-11
L1 # : 5340/66
M1 : 6146
N4 : 5621
N6 : 5040/42/43, 6154-56/59/62, 6412/14
Q4 : 5957/59, 6077, 6132/33/35-43/76/79

later reclassified L3

1st January 1928

C12 : 4534
C14 : 6123/25/27/28/30
J11 : 5210/18/22-24/39/88, 5306/17/26, 5982/90
M1 : 6146/49
N4 : 5513, 5620/21/25
N6 : 5040/43, 6154-56/59/62, 6412/14
O4 : 6246/49/57/61/96/98, 6305/12/26/55/73, 6523/24/26/27/39, 6585/88/95/97/98, 6638
Q4 : 5957/59, 6077, 6132-39/41-43/76

1st January 1932

C12 : 4515/34
C14 : 6123/25/27/28/30
J11 : 5222/35/39/49/88/99, 5300/06/12/26, 5982/90, 6044
M1 : 6147
N5 : 5923/26/39/46
N6 : 5040/41/43, 6154-56/58/59/62, 6412/14
O4 : 5384, 6246/49/57/96/98, 6305/26/44/73, 6523-25/37-39/85/88/97/98, 6605/34
Q4 : 5152, 5957/58, 6052/54/77, 6132-34/36-39/41/42/76/80

1st January 1936

J11 : 5218/22/49/88/99, 5300/06/12, 5952/74/82/90, 6044
J27 : 2360/61/89
M1 : 6147
N5 : 5526/42/46, 5894/98, 5903
N6 : 5041/43, 6155, 6412
O4 : 5384/98, 5400/07, 6206/96/98, 6305/26/44/73, 6522-24/37-39, 6585/88/97/98, 6605/34
Q4 : 5957/58, 6052/54/77, 6133/34/36/39/41/42/76/80

1st January 1940

C14 : 6124/31
J11 : 5206/18/23/34/49/50/52/84/88/99, 5300/03/06/08/12/22, 5952/74/82, 6044
J21 : 26, 300, 470, 806, 1806
N5 : 5526/42, 5746/62, 5894/95/99, 5903
O4 : 5335/84/98, 5407, 6206/07/25/66/71/77/96/98, 6305/26/44/56, 6505/24/37/39/85/97/98, 6605
Q4 : 5044/67, 6052/54, 6133/34/36/39/76/79

1st January 1944

D11 : 5501/03/11
J11 : 5206/18/23/49/50/99. 5300/03/06/08/12/13/22, 5952/74/78/82/83, 5986/90, 6003/44/81, 6115
N5 : 5746/65, 5894/95/99, 5903
O4 : 5335/84/98, 5406/07, 6205-07/40/71/77/90/95/96/98, 6305/07/09/20/26/28/44/56/71/74, 6505/11/12/19/39/44/50/53/73/91/98, 6619
Q1 : 5048/70

1st January 1948

			(pre-1946 numbers for comparison)
C12	:	7351/55/57/84	4010/16/19, 4531
J11	:	4281/89/97, 4321/33/58/78/79/89	5206/07/49/50/99, 5303/12/13
		4414/18/26/27/50	5952/74/82/90, 6044/81
N5	:	9284, 9319/23/27	5746, 5895/99, 5903
O2	:	3921/23-28/42/43/45/46/64-74/76-87	3461/78-83/97/98, 3500/01, 3834-44/46-57
O4	:	3597, 3615/44/51/65/77, 3837	5390, 5407, 6250, 6357, 6523 6539/43
Q1	:	9928/29	5048/70

1st January 1952 (6-prefix omitted throughout)

A5 : 9812/15/21
J11 : 4281/89/97, 4310/21/37/58/78/79/89, 4408/14/26/27
N5 : 9284, 9319/23/27
O4 : 3577/85/97, 3603/15/27/32/43/44/64/65/67/79/83, 3703/07/09/15/17/24/32/50/58/59/65/76, 3800/09/33/37 3840/42/70, 3902
Q1 : 9928/29

1st January 1956 (6-prefix omitted throughout)

J11 : 4281/93/97, 4310/37/79/89, 4408/14/27
N5 : 9284, 9319
O4 : 3577/85, 3632/34/36/64/67/79, 3703/07/15/17/32/39/65/76, 3800/33/40/42/61/70, 3902
Q1 : 9928/29
WD : 90043/51/55/87, 90162, 90259/75/87, 90302, 90411/31/49/92, 90502/45/54/67/77/94

1st January 1960 (6-prefix omitted throughout)

J11 : 4314/16/17/24/32/33/52/64/79, 4427
J67/69 : 8569/91
J94 : 8026/80
N5 : 9263/68/86/99
O4 : 3577/97, 3607/12/34-36/43/64/79/83/91/97, 3703/04/07/15/17/18/31/32/39/58/63/65/76, 3800/01/29/33/40/42/53/61/70/93, 3902/12
WD : 90043/88, 90259/75/87, 90302, 90431/38/49/92, 90545/54/77/94, 90657

19th June 1965

B1 : 61313/15/72/94
O4 : 63589/90, 63612/30/46/50/79/91/97, 63701/06/32/39/68, 63828/43/50/68/93
WD : 90043/69/88, 90149/53, 90271/75/92, 90301/98, 90401/18/49, 90558, 90658, 90719
43XXX : 43064
9F : 92039-42, 92141/44-46/48/49/73/78/79/82/86/89/91, 92200

(For Annual Summary, see pages 88/89)

Appendix 2 : Booked Departures ex Langwith Loco, Winter 1946

A.M.	Freq	To	Notes
12.15	D	Warsop Jct.	To work 1.03 to Whitemoor
12.40	MX	Warsop Jct.	To work 1.15 to Retford
12.45	D	Mansfield C.S.	To work 2.00 to New England
1.00 }	MO	Warsop Jct.	To work 1.30 to Mottram
1.00 }	MO	Warsop Jct.	To work 1.30 to Lincoln
2.15	MX	Mansfield C.S.	To work 3.20 to Mottram
2.45	D	Mansfield C.S.	To work 3.50 to Whitemoor
2.45	MO	Warsop Jct.	To work 3.17 to Whitemoor
3.18	MO	Warsop Jct.	To work 3.48 to Mottram
3.30	MO	Langwith Jct.	Langwith Jct. Pilot
3.30	D	Warsop Jct.	To work 4.00 to Whitemoor
3.35 }	MX	Mansfield C.S.	To work 4.51 to Hull
3.35 }	MX	Mansfield C.S.	To work 5.00 to Mottram
4.25	Q	Langwith Jct.	To assist 3.00 Tuxford -Attercliffe
5.25	MX	Warsop Jct.	Makes up and works 7.05 to Shirebrook (LMS)
5.25	MO	Warsop Jct.	To work 6.00 to Retford
5.40	MO	Langwith Jct.	To work 5.55 to Kirkby-in-Ashfield.
5.40 }	MX	Mansfield Station	Mansfield Station Pilot
5.40 }	D	Mansfield C.S.	Mansfield C.S. Pilot
5.40 }	MO	Mansfield C.S.	To work 6.55 to Mottram
5.55	D	Warsop Jct.	To work 6.25 to Sheffield
6.00	MO	Warsop Jct.	Warsop Jct Pilots (2 engines)
6.28	MO	Warsop Jct.	To work 7.05 to Shirebrook LMS
6.40	MO	Mansfield C.S.	To work 8.10 to Hull
6.45 }	D	Shirebrook North	To work 7.05 Pass. to Chesterfield
6.45 }	D	Shirebrook North	To work 7.25 Pass. to Lincoln

7.05	D	Markham Jct.	To work 8.40 to Frodingham
7.30	D	Warsop Jct.	To work 8.00 to Woodford
7.40	D	Markham Jct.	Markham Jct. Pilot
7.50	D	Warsop Jct.	To work 8.25 to Mottram
8.00	MX	Warsop Jct.	Warsop Jct. No. 1 Pilot
8.02	D	Warsop Jct.	To work 8.32 to Lincoln
8.17	D	Mansfield C.S.	To work 9.38 to New England
8.25	D	Warsop Jct.	To work 9.00 to Ickles
8.50	D	Warsop Jct.	To work 9.20 to Sherwood
8.50	Q	Warsop Jct.	To work 9.10 to Hull
8.55 }		Mansfield C.S.	To work 10.10 to Mottram
8.55 }	D	Mansfield C.S.	Rufford Colliery Pilot
9.15	D	Mansfield C.S.	To work 10.37 to Grimsby
11.05	D	Mansfield C.S.	To work 12.10 to Immingham
11.10	D	Warsop Jct.	To work 11.40 to Lincoln
11.35	D	Warsop Jct.	To work 12.05 to Broughton Lane (Sheffield)
11.50	D	Warsop Jct.	Warsop Jct. No. 2 Pilot

P.M.	Freq.	To	Notes
12.30	D	Mansfield C.S.	To work 1.45 to Whitemoor
12.50	D	Warsop Jct.	To work 1.20 to Staveley
1.20	D	Warsop Jct.	To work 1.58 to Mottram
2.00	D	Langwith Jct.	Langwith Jct. Pilot
2.45	D	Mansfield Station	To work 4.00 Pass. to Chesterfield
2.50	D	Warsop Jct.	To work 3.30 to Grimsby
4.02	TuThSO	Mansfield C.S.	To work 5.08 to Frodingham
4.18	MWFO	Warsop Jct.	To work 4.55 to Frodingham
4.50	D	Warsop Jct.	To work 5.22 to Woodford
5.08	D	Thoresby Colliery	To work 5.50 to Sheffield
6.25	D	Warsop Jct.	To work 6.57 to Tuxford
6.45	D	Chesterfield	To work 9.10 to Langwith Jct.
7.35	D	Warsop Jct.	To work 8.10 to Rotherham Rd.
8.30	D	Mansfield C.S.	To work 9.49 to Whitemoor
9.00	D	Mansfield C.S.	To work 9.25 to New England
10.10	SX	Mansfield C.S.	To work 11.15 to Mottram
10.40	D	Warsop Jct.	To work 11.20 to Whitemoor
11.50	SX	Warsop Jct.	To work 12.20 to Lincoln

Notes : Mansfield C.S. = Mansfield Concentration Sidings
D = Daily
Q = Conditional (i.e. runs when required)

Other Pilot Engines provided by Langwith Loco:-

Blidworth Colliery
Cresswell Colliery No.1 and No.2
Oxcroft Colliery
Sherwood Colliery
Shirebrook Colliery
Thoresby Colliery
Warsop Main Colliery
Warsop Sand Quarry
Welbeck Colliery

Appendix 3 : Booked Departures ex Langwith Loco, Winter 1962

A.M.	Freq.	To	Notes
12.10	MX	Mansfield C.S.	Mansfield C.S.Pilot
12.25	MX	Mansfield C.S.	To work 1.50 to Mottram
1.05	MX	Mansfield C.S.	To work 2.27 to Hull
1.05	MX	Mansfield C.S.	To work 2.20 to Colwick
4.00	D	Warsop Jct.	To work 4.40 to Aldwarke Main
4.25	MO	Warsop Jct.	To work 5.40 to Mottram
4.25	MX	Mansfield C.S.	To work 5.25 to Mottram
4.35	SX	Warsop Jct.	To work 6.13 to Northwich
4.35	MX	Warsop Jct.	Assists " " "
5.06	D	Mansfield C.S.	To work 6.20 to Annesley
5.20	D	Shirebrook West	To work 6.03 to Sheffield
5.20	D	Mansfield C.S.	
5.35	D	Warsop Jct.	To Warsop Main Colliery
5.35	MO	Warsop Jct.	To work 6.10 to Lincoln (Holmes)
5.45	MO	Mansfield C.S.	To work 7.07 to Colwick
5.45	MO	Mansfield C.S.	Mansfield C.S. Pilot
5.45	D	Mansfield C.S.	Mansfield Station Pilot
6.05	D	Warsop Jct.	To work 6.50 to Mottram
6.35	D	Mansfield C.S.	
6.35	D	Mansfield C.S.	
6.40	D	Welbeck Colliery Jct.	
6.50	D	Mansfield Colliery	
7.00	D	Ollerton Colliery	
7.10	D	Mansfield C.S.	
7.15	D	Warsop Jct.	To work 8.00 to Annesley
7.20	D	Thoresby Colliery Jct.	
8.05	D	Bevercotes Colliery	
8.05	Q	Ollerton Colliery	To work 9.12 to Eakring Oil Sidings
8.40	MO	Warsop Jct.	
9.15	MWFO	Warsop Jct.	To work 9.53 to Lincoln (Holmes)
9.45	SO	Welbeck Colliery Jct.	
9.50	D	Mansfield C.S.	
10.00	MSX	Mansfield C.S.	
10.15	SX	Warsop Jct.	To work 11.22 to Mottram
10.55	D	Mansfield C.S.	To work 12.02 to Mottram
11.13	MO	Welbeck Colliery Jct.	
11.38	SX	Warsop Jct.	

P.M.	Freq.	To	Notes
12.55	D	Warsop Jct.	SO to work 1.35 to Doncaster Decoy SX to work 1.35 to Hull
1.23	SO	Mansfield C.S.	To work 2.48 to Mottram
1.24	SX	Mansfield C.S.	assists 2.25 to Rufford
1.33	SX	Welbeck Colliery Jct.	
2.35	MO	Warsop Jct.	
3.05	SO	Warsop Jct.	To work 3.50 to Staveley

3.20	D	Mansfield C.S.	To work 4.45 to Annesley
3.35	SO	Mansfield C.S.	To work 4.58 to Hessle Haven
4.30	SO	Warsop Jct.	To work 5.40 to Mottram
4.40	SO	Mansfield C.S.	To work 6.08 to Mottram
8.21	SX	Mansfield C.S.	Rufford Stocking Site Pilot
8.27	SX	Mansfield C.S.	To work 10.15 to Northwich
9.35	TuWFO	Thoresby Colliery Jct.	
10.30	SX	Warsop Jct.	To work 11.10 to Lincoln (Holmes)

Other Pilot Engines provided by Langwith Loco:-

High Marnham No. 50 High Marnham No. 51 High Marnham No. 52 High Marnham No. 53	} to work between producing points and power station

Langwith Jct. No. 1	marshalls as required	
Mansfield C.S.No. 2	diesel shunter	
Warsop Jct. No. 1	" "	– Up side
Warsop Jct. No. 2	" "	– Down side

(58) *A grimy J11, 64289, at the east end of the shed on 12th March 1950. This "Pom -Pom" was shedded at Langwith continuously from before 1935 until withdrawal in June 1955.*

Appendix 4 : Passenger Train Departures - Langwith Junction (Shirebrook North)

1) 1906 (LD&ECR)

a) To Lincoln (Great Northern)
a.m. : 10.45
p.m. : 2.17 (FO), 2.17 (FX - to Ollerton only), 4.38, 7.20 (SO)

b) To Sheffield (Midland)

a.m. : 10.44
p.m. : 12.50 (SO), 2.17, 4.39, 5.30 (to Clowne only), 7.24, 9.00* (SO - to Clowne only), 9.25 (SO), 11.40* (SO - to Clowne only)

c) To Chesterfield (Market Place)
a.m. : 8.30, 10.46
p.m. : 12.05 (SO), 1.20 (SO), 2.07 (MThX), 2.17 (MThO), 3.00 (SO) 4.00 (SO), 4.40 (FX), 5.20 (FO), 6.10* (SO), 7.23, 9.00 (SO) 10.20 (SO)

d) To Mansfield (Midland)
a.m. : 10.41
p.m. : 2.05, 4.33, 7.24

* These trains ran between Clowne and Chesterfield, reversing at Langwith Jct.

Note that at this period, the 'main line' was regarded as Sheffield to Lincoln via Langwith Junction, with the Chesterfield line treated as a branch.

It can also be seen that even at this early date, the tradition of four trains connecting at Langwith Junction several times each day, had already been established.

2) April 1910 (GCR)

a) To Lincoln
a.m. : 10.45
p.m. : 2.17, 4.06 (to Ollerton only), 4.38, 6.53 (conveys through coach from Marylebone, Leicester and Nottingham via Heath), 7.20 (to Ollerton only).

b) To Sheffield
a.m. : 10.45
p.m. : 1.12 (SO), 2.17, 4.39, 5.30 (SO - to Clowne only), 7.25, 9.05 (SO), 11.45 (SO - to Clowne only)

c) To Chesterfield
a.m. : 8.30, 9.16 (to Bolsover only, then via Heath and G.C. Main Line to Marylebone), 10.46
p.m. : 1.20 (SO), 2.17, 3.30 (SO), 4.40 (FX), 5.23 (FO), 6.32 (SO) 7.23, 9.00 (SO)

d) To Mansfield (Midland)
a.m. : 10.41
p.m. : 2.12, 4.33, 7.23, 11.47 (SO)

Note the through service from London to Lincoln via the LD&EC line. The evening service was a Slip Carriage, detached from the 3.15 p.m. ex Marylebone at Leicester, worked forward via Nottingham and Heath, thence over the connection at Duckmanton Jct. to Langwith Jct. where it was attached to the 6.53 p.m. all-stations to Lincoln, arriving there at 7.58 p.m.

The morning service left Lincoln at 8.10 a.m. attached to the all-stations local to Langwith Jct., arriving there at 9.15 a.m. It was detached there and worked forward by the reverse route to that described above, calling at Heath (where there was a connection to Sheffield). At Nottingham it was attached to the "Breakfast & Luncheon Corridor Express" ex Sheffield, which despite its grand title called at such places as Finmere and Calvert before arriving at Marylebone at 1.30 p.m.

The concurrent journey time from Lincoln to London via Grantham and the Great Northern was about three hours!

Rather curiously, the 1910 timetable lists Clowne under its original spelling of "Clown".

3) July 1938 (LNER)

a) To Lincoln
   a.m. : 10.46
   p.m. : 1.54 (FO), 4.30, 7.30 (SO)

b) To Sheffield
   a.m. : 10.43 (SX), 10.48 (SO),
   p.m. : 1.10 (SO), 4.36, 7.30

c) To Chesterfield
   a.m. : 7.15, 8.27, 10.46
   p.m. : 1.15 (SO), 1.43, 2.13 (SO), 4.35, 5.52 (SO), 6.31 (SX), 7.30, 8.50 (SX), 9.20 (SO), 10.25 (SO), 12.00 (SO - to Bolsover only)

d) To Mansfield (LMS)
   a.m. : 10.40
   p.m. : 4.27, 7.28, 12.02 (midnight)(SO)

e) To Mansfield (LNER), via Warsop, with connections to Nottingham*
   a.m. : 6.30
   p.m. : 1.37, 2.57 (SO), 10.26 (SO)

* The direct service from Langwith Jct. to Nottingham via Shirebrook South and the Leen Valley line had only run from 1925-31.

There were never any regular Sunday passenger trains on the line.

Appendix 5 : A Rainy Day at Langwith Loco

The weather was appropriately gloomy, albeit unseasonal, for what turned out to be my final *official* visit to the Loco Depot, which took place on Sunday 27th May 1962 in company with a party of RCTS members from Bristol.

Time was obviously running out for the old steam shed and its occupants; eight diesel locos stood among the total of 62 engines present, a total well down on previous years, when upwards of a hundred engines would be on shed each Sunday.

The shed allocation plate on each smokebox door was recorded, indicating that of the 54 steam locos 35 were residents, 14 were visitors and the remainder unmarked. Not too much reliance can be placed on these figures, however, as by this date steam was rapidly being run down and updating shed plates didn't figure very highly in importance. It seems likely that most of the 'visitors' would have been Langwith residents.

The mix of engines present showed some differences from that found on earlier visits. The passenger tanks had of course long gone, being transferred away in 1955 soon after the Lincoln service was withdrawn, and the last two N5's at Langwith, 69263 and 69286, had been scrapped in November 1960, almost the last of their type. Their supplanters, the J94 0-6-0ST's, were in turn being replaced by 350hp. diesel shunters. The old faithful O4 2-8-0's were still present in some numbers, supplemented by a dozen WD's. K3 61811, the only one of its class to run with an ex-V2 tender, was under repair in the Carriage Shed, which also housed four WD's and the Breakdown Train.

(59) *"The New Order Cometh" : Class J94 0-6-0ST 68020, itself a replacement for the old N5 shunting engines, stands in the shed yard on 27th May 1962 with evidence of its own demise - diesel power!*

The long siding in the Down Yard immediately alongside the Leen Valley running lines was as usual full of engines - two B1's (61044 and 61152) and J94 68078 were stored by the bufferstop, next were fifteen O4's, which included 63720 which appeared to be recently ex-works, 63577 in poor condition with a damaged o/s cylinder, and three more (63731, 63853, 63902) all displaying evidence of a hard life with drooping front ends.

Near the turntable stood the Depot Snowplough DE330917, mounted on an old GNR tender still carrying its number plate 5131. On the back road formerly used for coal-stocking, J11 64379 was supplying steam to the Instruction Train, a set of ancient coaches converted into classrooms and mobile workshops; 64379 was a long-time resident of Langwith, having arrived from Colwick in 1939 and destined to be Langwiths last J11 when withdrawn four months after this visit.

(60) *A hum-drum duty for an old-stager; Langwith's last J11, 64379, supplies steam heating for the visiting Instruction Train, parked on the old coal-stacking road at the Loco, 27th May 1962.*

There was an echo from the past during the visit, when a Diversion appeared in the shape of B1 61327 at the head of an Up express re-routed off the Great Central main line due to engineering work.

Summary of locomotives present :

B1	:	61043/44, 61152, 61377/99
K3	:	61811/40/86, 61946
O4	:	63577, 63664/79/83/87/97, 63703/15/17/20/22/31/39/63/65/74/76, 63800/01/03/28/29/40/42/53/61, 63902/12
J11	:	64318/79
J94	:	68020/69/78
WD	:	90000/043/145/275/301/302/411/418/431/449/657/660
350hp shunters	:	D3701, D4053/60/66/67/69
Brush Type 2	:	D5814/28

Total 62

Appendix 6 : Langwith Loco - The Final Year

1965 proved to be the last full year of operation at the Loco; the following notes were made during my intermittent visits in this period.

Christmas Day 1964

In anticipation of the snow which began to fall on Boxing Day, two locomotives were left in steam for snowplough duty - O4 63717 fitted with a small bufferbeam plough, and WD 90301 coupled to the tender plough. If required they would work back-to-back with the WD leading.

Waiting for their final journey to the breakers yard were O4's 63615/36 and 63800 along with WD 90209. Under repair were 63697 and 63907, whilst other O4's still present comprised 63679/83/91, 63703/32/39, 63842/43/50/61/77/82/93 and 63902. The remainder of the dwindling steam allocation at this time consisted of WD 2-8-0's, of which 90043/88, 90121/64, 90227/58/66/71/75/92, 90398, 90418/49 and 90587 were on shed. A sign of the times was the presence of a new Type 1 Bo-Bo D8608, its shining appearance in marked contrast to that of the steam locomotives and the grubby diesel shunters also present.

(On the same day, the still under-construction diesel depot at Shirebrook West contained five Brush Type 4's and another 350hp shunter).

March 13th 1965

By this date nineteen 9F 2-10-0's had arrived from the closed shed at Peterborough (New England), along with some LMS-type 43xxx 2-6-0's. 63615 was still in store, and had been joined by 63683 and 63861. Five other O4's were in steam or showed signs of recent use.

(61) *The Final Year : Class O4/8 63691 stands in the yard on 22nd May 1965, less than a month before withdrawal. The Engine Shed has received a new roof, but the old Carriage Shed seems in a parlous state!*

(62) *A 'Space-ship' under the 'Cenotaph'. 92178 takes coal April 1965 during the brief period when the 9F's monopolised local coal traffic.*

(63) *9F 2-10-0 92042 works an Up mixed freight off the Beighton line through Platform 4, 20th November 1965; judging by its appearance it is no surprise to learn that the 'Space-ship' was withdrawn the following month. Note that the former west crossover which had given direct access to Newtons Sidings on the right has been lifted; surplus wagons are stored in Platform 2.*

(64) *Class O4/7 2-8-0 63843 receives some fairly major attention under the sheerlegs, 22nd May 1965; she has only six months life left before withdrawal.*

September 16th 1965

By now the number of serviceable O4's had dwindled to four, including 63612 which left for a trip working, and 63739 which was on the Langwith Colliery -Warsop Jct duty. 9F 92200 was seen attached to the Breakdown Train, on what might have been its final duty, as it was withdrawn shortly afterwards; by November it was photographed being broken up at Ward's Scrapyard, Killamarsh - it was just *five years old!*

The Carriage Shed was empty apart from three O1's , 63589, 63768 and 63868, all with dismantled motion awaiting their last journey. Laid off in the yard were several WD's and O4's 63697, 63732 and 63828.

Christmas Day 1965

This proved to be my last visit to the Loco whilst it was still in use. The Running Shed sheltered B1's 61051 and 61394, the latter already withdrawn; 61051 would only run for a week or two before being transferred briefly to Stationary Boiler Stock. Also in the Shed were eight WD's, 90069/149/153/401/418/449/572/719.

The Carriage Shed housed the last two O4's at Langwith, 63612 and 63843, withdrawn in November and due to leave for Cox & Danks scrapyard at Wadsley Bridge on January 7th. 63612 had the melancholy distinction of being the last surviving Great Central-built locomotive in service on B.R. Alongside were more withdrawn engines - WD's 90220/529/587/697 and 2-6-0 43149, the last of its type on the Eastern Region.

Brush Type 2 D5686 was on snowplough standby, with old Great Northern tender ploughs tethered either end - the Langwith plough DE330917 and a larger version, 966, recently transferred from Staveley.

Dumped around the yard were more withdrawn locos; WD 90558 near the turntable, 90189/551/658 on the ashpit road and 90043, along with Langwiths last 9F 92039 and two more 2-6-0's, 43082 and 43109, on the Back Road near the coaling plant.

- - - - - - - -

For the record, the last WD's withdrawn from Langwith were :

January 1966 : 90069, 90149, 90258, 90418/449*
February 1966: 90153, 90572, 90719

(* 90449 was the first WD to be transferred to Langwith, in June 1954)

- - - - - - - -

The very last normal steam working noted at Langwith Junction was in July 1966, when WD 90254 ran light down the Leen Valley line and on to Sheffield.

Final steam appearance of any kind occurred in 1968, when "Flying Scotsman" stopped in Platform 4 for water whilst hauling a Great Central Rail Tour.

- - - - - -

Possibly the last working of any significance at the Junction was a ten-coach railtour which struggled up the "New Found Out" on 24th November 1990, "top-and-tailed" by Brush Class 4's 47489 and 47844. It was a dull, misty, funereal sort of day; perhaps the Junction was mourning its past glories!

- - - - - - - - -

(65) *Langwith Junction today; a very short length of Platform 4 survives alongside the bufferstop at the end of Davis's headshunt; the track was formerly part of the Down Beighton line, and the discarded oildrum marks the site of the signalbox.*

Summary

1st January : Class	1923	1928	1932	1936	1940	1944	1948	1949	1950	1951	1952	1953	1954
Passenger													
A5	-	-	-	-	-	-	-	-	3	3	3	3	3
C12	-	1	2	-	-	-	4	2	-	-	-	-	-
C13	-	-	-	-	-	-	-	-	-	-	-	-	-
C14	3	5	5	-	2	-	-	-	-	-	-	-	-
D11	-	-	-	-	-	3	-	-	-	-	-	-	-
G3 *	6	-	-	-	-	-	-	-	-	-	-	-	-
Freight													
J11	16	12	13	13	20	24	14	14	12	15	14	15	15
J21	-	-	-	-	5	-	-	-	-	-	-	-	-
J27	-	-	-	3	-	-	-	-	-	-	-	-	-
L1(L3)	2	-	-	-	-	-	-	-	-	-	-	-	-
M1 *	1	2	1	1	-	-	-	-	-	-	-	-	-
O2	-	-	-	-	-	-	34	33	33	-	-	-	-
O4	-	22	22	23	24	37	7	7	9	39	34	33	34
Q1	-	-	-	-	-	2	2	2	2	2	2	2	2
Q4	16	15	17	13	10	-	-	-	-	-	-	-	-
WD	-	-	-	-	-	-	-	-	-	-	-	-	-
Mixed Traffic													
B1	-	-	-	-	-	-	-	-	-	-	-	-	-
K3	-	-	-	-	-	-	-	-	-	-	-	-	-
Shunting													
J58	1	-	-	-	-	-	-	-	-	-	-	-	-
J60 *	4	-	-	-	-	-	-	-	-	-	-	-	-
J69	-	-	-	-	-	-	-	-	-	-	-	-	-
J94	-	-	-	-	-	-	-	-	-	-	-	-	-
N4	1	4	-	-	-	-	-	-	-	-	-	-	-
N5	-	-	4	6	8	6	4	4	4	4	4	3	3
N6 *	10	9	11	4	-	-	-	-	-	-	-	-	-
08 (diesel)	-	-	-	-	-	-	-	-	-	-	-	-	-
Total	60	70	75	63	69	72	65	62	63	63	57	56	57

* original LD&ECR locomotives

Class	1955	1956	1957	1958	1959	1960	1961	1962	1963	1964	1965	1966
Passenger												
A5	-	-	-	-	-	-	-	-	-	-	-	-
C12	-	-	-	-	-	-	-	-	-	-	-	-
C13	2	-	-	-	-	-	-	-	-	-	-	-
C14	-	-	-	-	-	-	-	-	-	-	-	-
D11	-	-	-	-	-	-	-	-	-	-	-	-
G3	-	-	-	-	-	-	-	-	-	-	-	-
Freight												
J11	12	10	10	9	10	10	10	3	-	-	-	-
J21	-	-	-	-	-	-	-	-	-	-	-	-
J27	-	-	-	-	-	-	-	-	-	-	-	-
L1(L3)	-	-	-	-	-	-	-	-	-	-	-	-
M1	-	-	-	-	-	-	-	-	-	-	-	-
O2	-	-	-	-	-	-	-	-	-	-	-	-
O4	29	23	23	23	23	38	31	34	26	22	17	-
Q1	2	2	2	2	-	-	-	-	-	-	-	-
Q4	-	-	-	-	-	-	-	-	-	-	-	-
WD	12	19	18	21	18	15	9	11	11	14	16	8
Mixed Freight												
B1	-	-	-	-	-	-	1	1	-	-	-	3
K3	-	-	-	-	-	-	-	1	-	-	-	-
Shunting												
J58	-	-	-	-	-	-	-	-	-	-	-	-
J60	-	-	-	-	-	-	-	-	-	-	-	-
J69	-	-	-	-	-	2	1	-	-	-	-	-
J94	-	-	-	-	2	2	2	3	3	-	-	-
N4	-	-	-	-	-	-	-	-	-	-	-	-
N5	2	2	2	1	1	4	-	-	-	-	-	-
N6	-	-	-	-	-	-	-	-	-	-	-	-
08(diesel)	-	-	-	-	-	-	-	-	9	10	10	10
Total	59	56	55	56	54	71	54	53	49	46	43	21

n.b. certain types (9F 2-10-0, O1 2-8-0, LMR 43xxx 2-6-0, etc.), not shown above due to arriving after 1st January and departing before end of same year.

LANGWITH JUNCTION GOODS SHED
WOOD DOOR
ROAD SIDE ELEVATION
RAIL SIDE
BRICK
ROAD SIDE
END ELEVATION
BOTH IDENTICAL
RDAJ 10/94.
0 1 2 3 4 5 6 7 8 9 10 11 12 13 14 15 16 17 18 19 20 21 22 23 24 25 26 27 28 29 30 FEET
HALF PLAN·RAIL SIDE
SLIDING DOORS
NOTES
ROOF:
RIDGE TILES 12" LONG ON BLUE SLATE.
BRICKWORK:
RED BRICK WITH BLACK MORTAR.
BOTTOM EIGHT COURSES BLUE BRICK INCLUDING PILLARS, TOP ROW BEVELLED ON PILLARS. TWO TOP ROWS ARE BEVELLED BETWEEN PILLARS.
ROAD SIDE:
BLUE BRICK BEVELLED INNER EDGES TO PILLARS EITHER SIDE OF DOORS TO 84".
RAIL SIDE:
BLUE BRICK BEVELLED EDGE TO CENTRE FOUR PILLARS, BOTH EDGES TO 84".
↓ SINGLE DOWN PIPE.
DOORS:
TWIN PAIRS SLIDING DOORS. WOODEN PLANKS 5½" WIDE.
CANOPY SUPPORTS:
SIX ON RAIL SIDE, TWO ON ROAD SIDE.
MEASURED: L LITTLE, 7/78
DRAWN: RDA JOHNSTON. 10/94

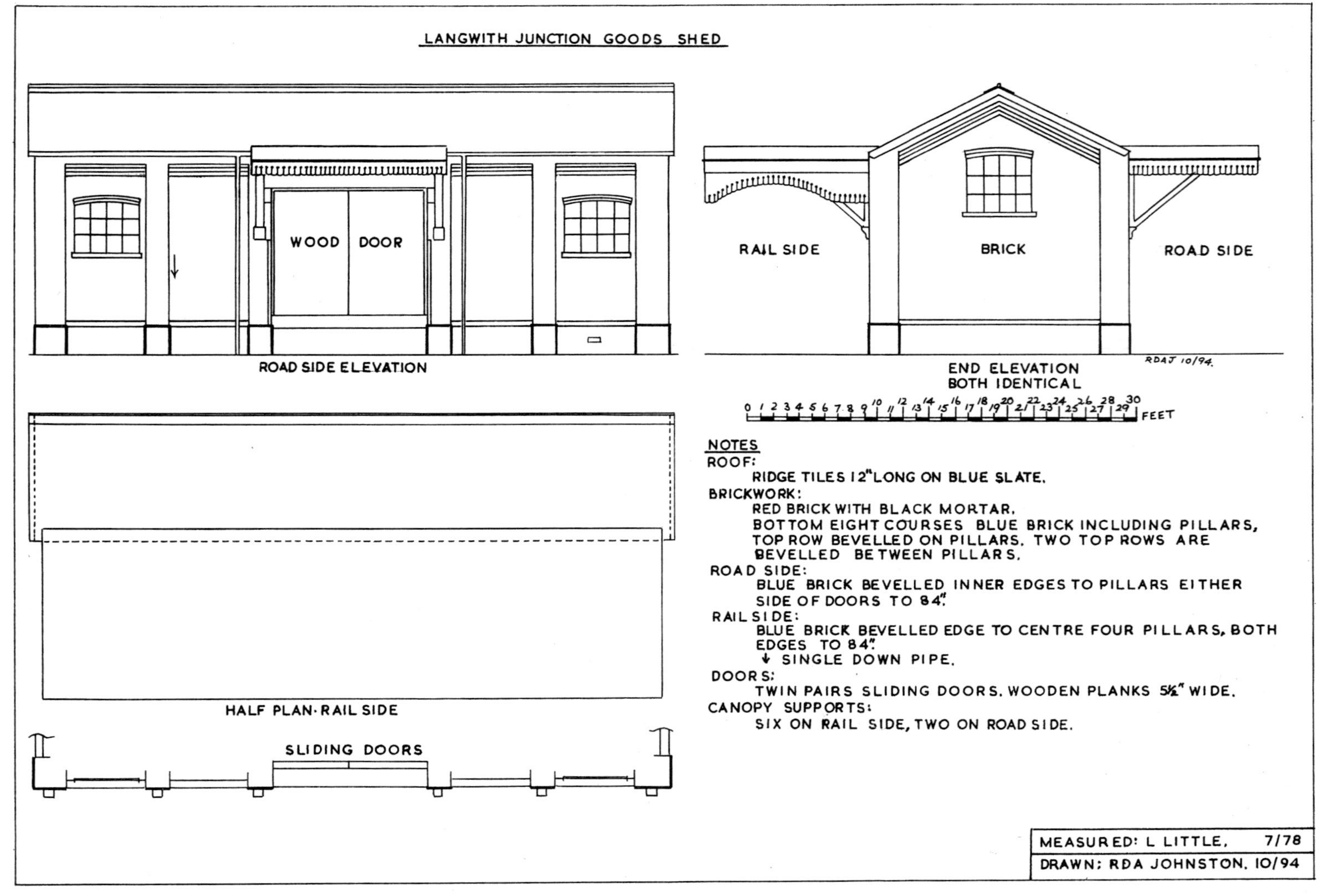

DIA.5

SAXBY & FARMER.Co - TAPPET LOCKING FRAME.
ILLUMINATED DIAGRAM
(EXLNE COMBINED INSTRUMENT)
PLEASLEY (LEEN VALLEY)
UP NON PEG
DN PEG
WARSOP JCT. (MAIN LINE)
UP NON PEG
DN PEG
BEIGHTON BRANCH
LANGWITH COLLY. JCT.
UP PEG
DN NON-PEG
1 CK
A.B.C.D.I.
37, 38.
81.
NUMBERED 1 TO 81.
LANGWITH JUNCTION.- BLOCK SHELF/FRAME LAYOUT.-
FRONT ELEVATION. (NOT TO SCALE.)

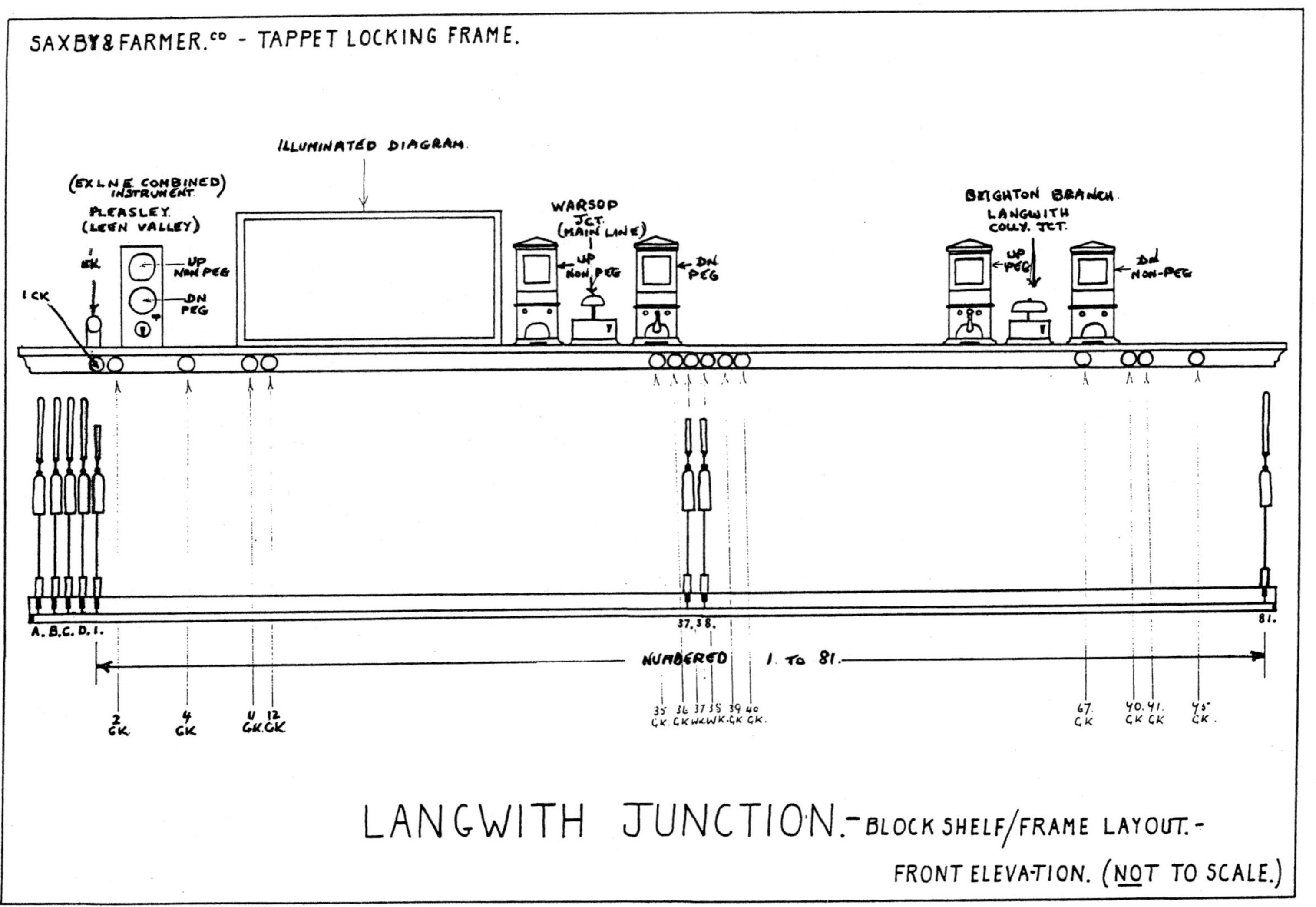

DIA.6

FRAME.
FOOTBOARD.
NUMBERED
1. TO 81.
A B C D
OUTER WALL
REMOVED. 1962.
SINK.
BRICK (LOWER).
LOWER
ROOM
DOOR.
LANDING.
TABLE.
LOCKER.
STOVE.
DESK
CLOCK.
PHONE
PHONE.
SHELF.
CONTROL
PHONE.
STOVE.
NOTE: MAJORITY
OF REAR WINDOWS
BLANKED OUT.
WHEELS.
NOTE:
RELAY ROOM
ADJOINING THIS
END OF BOX NOT
SHOWN.
LANGWITH JUNCTION.- PLAN. (NOT TO SCALE.)

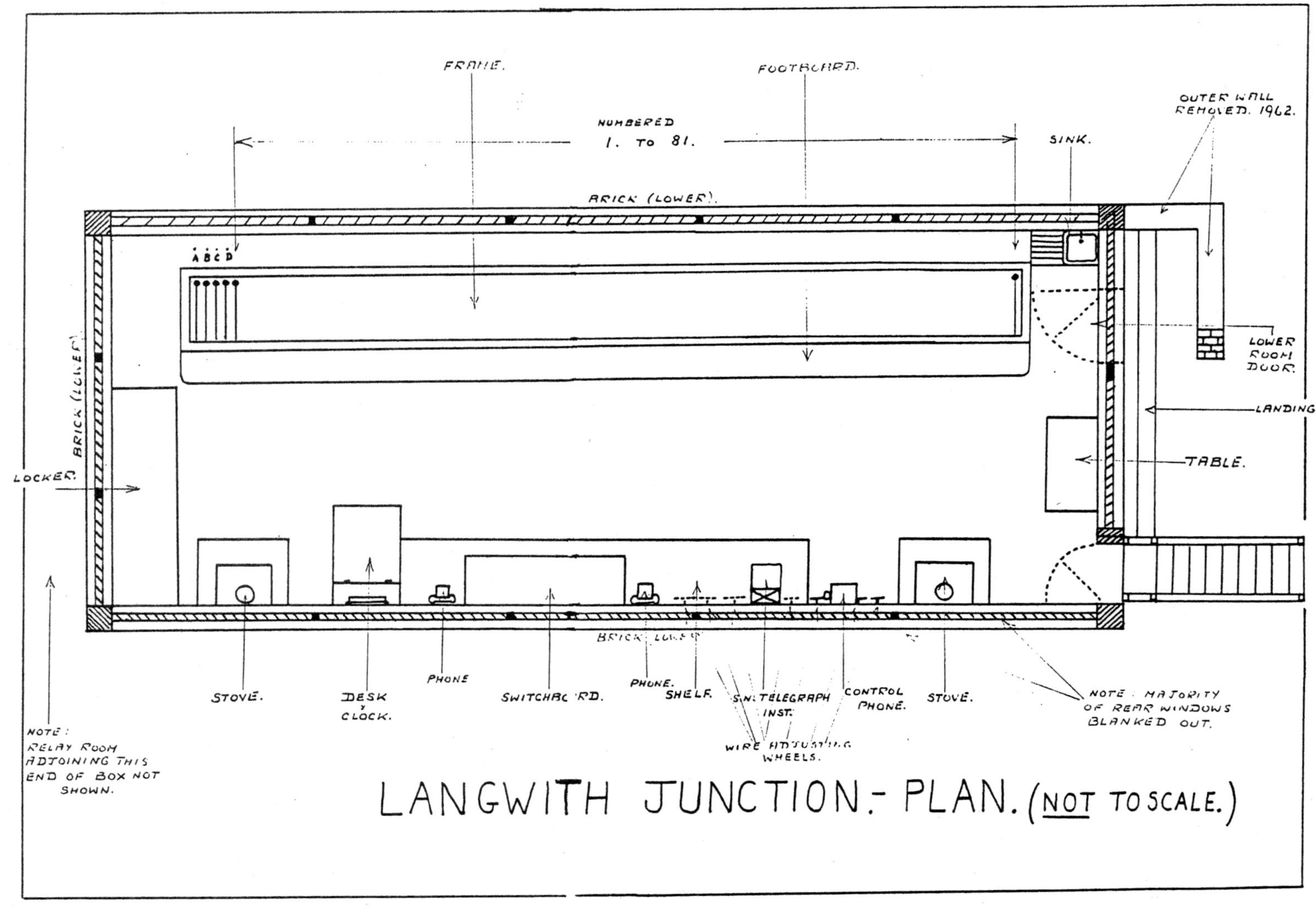

DIA.7